A ROAD MAP FOR SCALING UP PRIVATE SECTOR FINANCING FOR THE BLUE ECONOMY IN THAILAND

INVESTMENT REPORT

NOVEMBER 2023

ACGF
ASEAN CATALYTIC GREEN FINANCE FACILITY

ADB

A ROAD MAP FOR SCALING UP PRIVATE SECTOR FINANCING FOR THE BLUE ECONOMY IN THAILAND

INVESTMENT REPORT

NOVEMBER 2023

ACGF
ASEAN CATALYTIC GREEN FINANCE FACILITY

ADB

Note:
In this publication, "$" refers to United States dollars and "B" refers to Thailand baht.

ADB recognizes "China" as People's Republic of China and "Vietnam" as Viet Nam.

All photos are from ADB.

Cover: A cargo ship docking at Danang Port. The Port is the third largest port system in Viet Nam and lies at the eastern end of the GMS East–West Economic Corridor (EWEC), which connects Viet Nam with the Lao PDR, Thailand, and Myanmar.

Cover design by Ross Locsin Laccay.

Printed on recycled paper

CONTENTS

TABLE, FIGURES, AND BOXES

Table

Figures

Boxes

ABBREVIATIONS

ADB	Asian Development Bank
ACGF	ASEAN Catalytic Green Finance Facility
ASEAN	Association of Southeast Asian Nations
BCG	bio–circular–green
BOI Thailand	Thailand Board of Investment
BOT	Bank of Thailand
DMCR	Department of Marine and Coastal Resources
EEC	Eastern Economic Corridor
EGAT	Electricity Generating Authority of Thailand
ESG	environmental, social, and governance
FI	financial institution
GDP	gross domestic product
GHG	greenhouse gas
IUU	illlegal, unreported, and unregulated
MNRE	Ministry of Natural Resources and Environment
NPS	non-point source
OSMEP	Office of SMEs Promotion
PET	polyethylene terephthalate
R&D	research and development
SFI	specialized financial institution
SLL	sustainability-linked loan
SMEs	small and medium-sized enterprises
TCFD	Task Force on Climate-related Financial Disclosures
TCG	Thai Credit Guarantee Corporation

WEIGHTS AND MEASURES

GW	gigawatt
km	kilometer
MW	megawatt
Rai	4 Ngan (1,600 square meter)
TEU	twenty-foot equivalent unit

1 US dollar ($) = 35.1 Thailand baht (B) (2022)

ABOUT THIS INVESTMENT REPORT, THE AUTHORS, AND METHODOLOGY

This investment report supports the efforts of stakeholders to scale up financing from financial institutions toward blue economy sectors in Thailand. It details the importance of the blue economy and identifies investment opportunities, the barriers that hinder financing of such opportunities, and potential solutions to crowd in public and private capital. This report has been drafted for financiers, private sector stakeholders, policymakers, and development professionals in Thailand and the Southeast Asia region who are interested to develop and finance projects to support the growth of the blue economy.

The blue economy is defined as the sustainable use of ocean and coastal resources to drive economic growth and improve livelihoods while protecting and nurturing healthy marine ecosystems (Ecosystems [World Bank 2017]). Based on the Ocean Finance Framework of the Asian Development Bank (ADB), which is aligned with the United Nations (UN) Sustainable Blue Economy Principles, this report classifies 10 blue economy subsectors into three focus areas: (i) sustainable coastal development, (ii) pollution control, and (iii) ecosystem and natural resources management (ADB n.d., United Nations Environment Programme n.d.). As a coastal nation, Thailand derives significant socioeconomic value from its ocean-based industries. Therefore, ensuring the sustainable use of the ocean's natural assets is critical for safeguarding Thailand's future economic growth.

This investment report is based on the research and a more detailed report developed by AWR Lloyd and funded by ADB TA 9621 as part of its contribution to the ASEAN Catalytic Green Finance Facility. The report was prepared by Sree Kartha, ACGF consultant, under the guidance of Scott Roberts, head, Southeast Asia Green Finance Hub and the manager of ACGF, with inputs from Katharine Thoday, principal environment specialist (Environment Thematic Group Climate Change and Sustainable Development Department); Susan Olsen, senior investment specialist (Private Sector Operations Department); Ghislain De Valon, senior infrastructure specialist (Innovation and Green Finance); and Naeeda Crishna Morgado, senior infrastructure specialist (Innovation and Green Finance). The coordination and production of the investment report was managed by ACGF consultant Marina Lopez Andrich. This note was edited by Layla Yasmin Tanjutco-Amar and designed by Rocilyn Locsin Laccay.

The original report was developed by Ramadhan Putera Djaffri, Hyuna Jung, Leena Tricharoenwanich, and Natthida Luthra of AWR Lloyd.
AWR Lloyd reviewed publicly available information to establish a baseline of public- or private-led blue economy initiatives and identify all central government agencies that oversee Thailand's blue economy sectors. This research was supplemented with a review of Thailand's existing regulations and legislation related to the blue economy to identify relevant supervisory bodies. AWR Lloyd also conducted a comprehensive review of the established baseline to identify any non-ministerial level authorities that have been active in blue economy sectors. To map the interest of Thailand's financial institutions (including commercial banks and specialized financial institutions) in blue economy sectors, AWR Lloyd conducted interviews and extracted information from publicly available disclosures (including annual reports, sustainability reports, press releases, and news articles). AWR Lloyd also reviewed research reports, academic journals, and government publications and databases, and conducted interviews with key stakeholders such as financial institutions, private sector companies, government agencies, and sustainable finance experts to identify the barriers and potential solutions to scale up private sector financing for Thailand's blue economy.

A boater repaints his boat at
the riverbank of the Mekong
in Chiang Khong.

EXECUTIVE SUMMARY

This report develops and presents a road map for scaling up private sector financing toward the blue economy in Thailand. The road map categorizes blue economy sectors into four different groups with varying prioritization levels and action plans: hidden gems, question marks, quick wins, and stepping stones. The road map was prepared by assessing the attractiveness of blue economy sectors and the interest from financial institutions (FI) in these sectors.

PRIORITIZATION MATRIX OF BLUE ECONOMY SECTORS

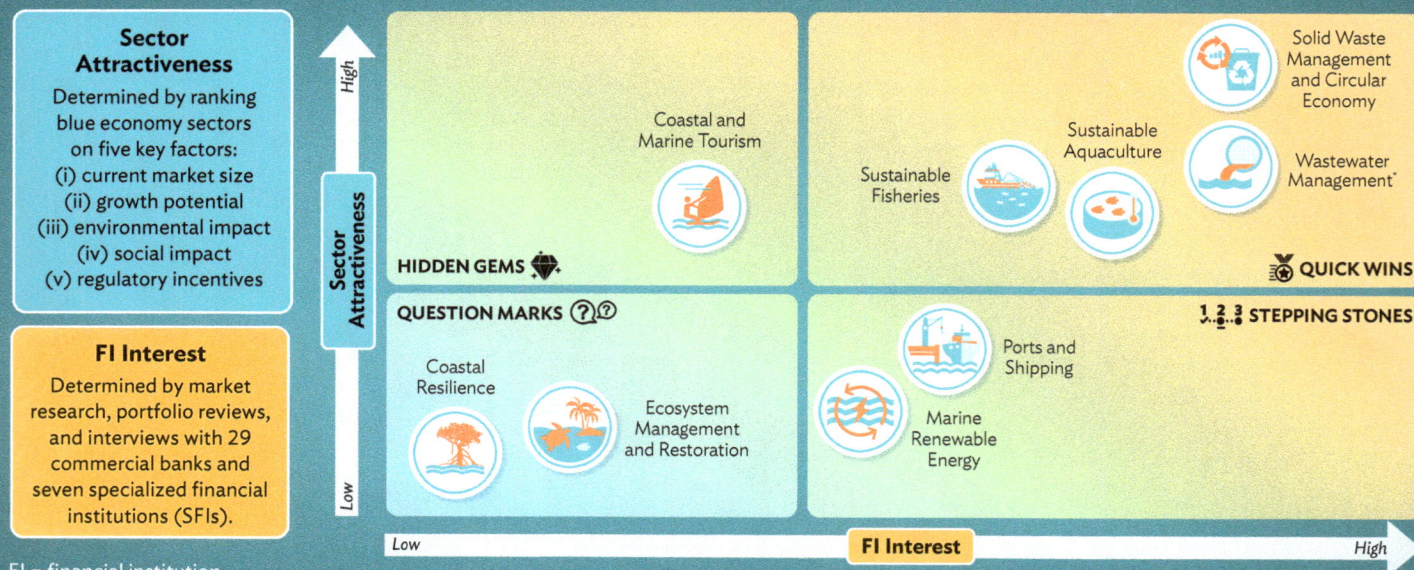

Sector Attractiveness
Determined by ranking blue economy sectors on five key factors:
(i) current market size
(ii) growth potential
(iii) environmental impact
(iv) social impact
(v) regulatory incentives

FI Interest
Determined by market research, portfolio reviews, and interviews with 29 commercial banks and seven specialized financial institutions (SFIs).

Sector Attractiveness (High ↑ / Low ↓) vs **FI Interest** (Low → High)

HIDDEN GEMS 💎
- Coastal and Marine Tourism

QUICK WINS
- Sustainable Fisheries
- Sustainable Aquaculture
- Solid Waste Management and Circular Economy
- Wastewater Management*

QUESTION MARKS ❓
- Coastal Resilience
- Ecosystem Management and Restoration

STEPPING STONES
- Marine Renewable Energy
- Ports and Shipping

FI = financial institution.
* Includes Non-point Source Pollution Management sector, which FIs usually incorporate in investments in Wastewater Management sector.
Source: AWR Lloyd.

Main Barriers

1. **Limited awareness** on the strategic importance of blue economy sectors for sustainable development.

2. **Slow pace of integration of sustainability targets** into core business activities for lenders and borrowers.

3. **Lack of monitoring and environmental, social, and governance compliance tools,** which contributes to the increased risk profile of financing such projects.

4. **Lack of investment readiness,** which presents challenges in attracting institutional capital.

5. **Limited availability of at-scale projects** also contributes to financing constraints.

6. **Small and medium-sized enterprise (SMEs)-related barriers** such as lack of collateral and credit history, which leads to increased perceived risk.

7. **Climate risks** increase perceived risk and limit the investible universe of projects.

Source: ADB based on AWR Lloyd.

SOLUTIONS

Recommended Product Designs

1. **Sustainability-linked Loans.** Use of proceeds is fungible, offering borrowers flexibility; incentivizes early adopters through dynamic pricing scheme with interest rates stepping down as borrowers meet pre-agreed sustainability targets.

2. **Guarantees.** First-loss or minimum take-or-pay guarantees mitigate against default risk and can help catalyze commercial capital; encourages commercial banks to lend to SMEs with limited collateral and credit history.

3. **Technical Assistance Programs.** Facilitates preparation, financing, and execution of blue economy projects by strengthening capacities and promoting resource efficiency of borrowers.

4. **Blue Carbon Finance.** Generates new revenue streams from the sale of carbon credits, thereby improving the risk profile of blue economy projects with strong public interest but lacking an established commercial business model.

5. **Community-based Revolving Loans.** By establishing revolving loan funds with local cooperatives, commercial banks can scale up SMEs loan portfolio costs effectively by disbursing and monitoring loans through partner cooperatives.

6. **Insurance.** Financiers can explore collaborations with insurers to develop insurance products that mitigate against physical risks of climate change facing blue economy projects and create tools to calculate risk exposure.

Recommended Policy Actions and Incentives

1. **Develop a Sustainable Finance National Taxonomy.** Creating a common national language covering all blue economy sectors will improve market clarity and lead to increased awareness and better integration of sustainability initiatives into core business activities.

2. **Create and Nurture a Better Data Environment.** A robust, reliable, and accessible data environment will enable companies to accurately measure and track sustainability metrics and comply with disclosure requirements of globally accepted sustainability frameworks.

3. **Promote Private Participation through Financial Incentives.** This could be achieved by redirecting unsustainable subsidies toward the blue economy, extending current pilot feed-in-tariff schemes toward marine renewable energy sources, creating tax holidays with corporate income tax exemptions or import duty waivers for projects related to the blue economy.

4. **Integrate Public Sector Efforts.** The blue economy cuts across multiple sectors and requires cohesive policies and effective networks to attract desired capital flows and to develop sustainable projects.

ROAD MAP FOR SCALING UP PRIVATE SECTOR FINANCING FOR THE BLUE ECONOMY IN THAILAND

Barriers | **SOLUTIONS** — $ Product Design | Policy Actions and Incentives

Quadrant 1: QUICK WINS

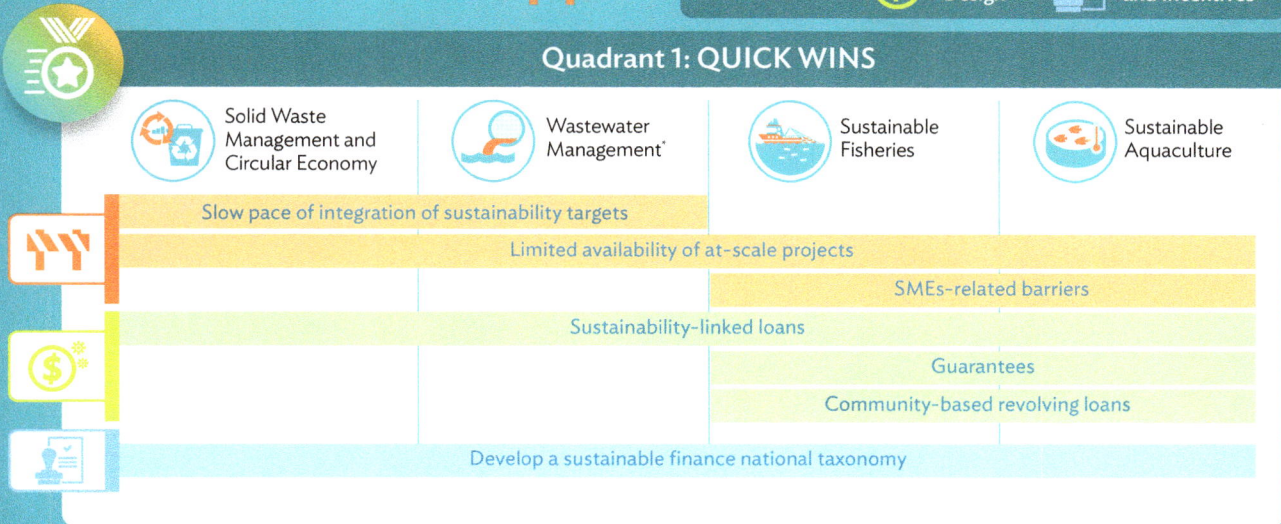

	Solid Waste Management and Circular Economy	Wastewater Management*	Sustainable Fisheries	Sustainable Aquaculture
Barriers	Slow pace of integration of sustainability targets			
	Limited availability of at-scale projects			
			SMEs-related barriers	
Product Design	Sustainability-linked loans			
			Guarantees	
			Community-based revolving loans	
Policy Actions	Develop a sustainable finance national taxonomy			

* Includes Non-point Source Pollution Management sector, which FIs usually incorporate in investments in Wastewater Management sector.

Quadrant 2: HIDDEN GEMS

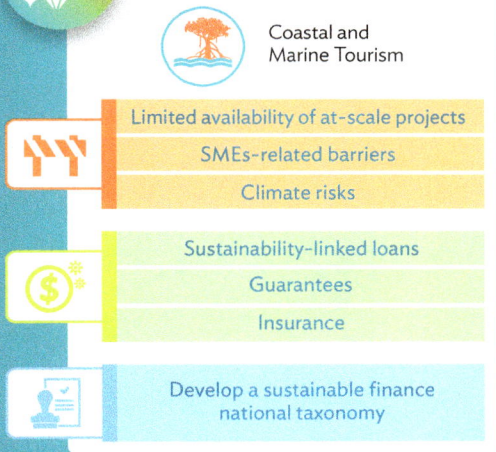

	Coastal and Marine Tourism
Barriers	Limited availability of at-scale projects
	SMEs-related barriers
	Climate risks
Product Design	Sustainability-linked loans
	Guarantees
	Insurance
Policy Actions	Develop a sustainable finance national taxonomy

Quadrant 3: STEPPING STONES

	Ports and Shipping	Marine Renewable Energy
Barriers	Lack of monitoring and ESG compliance tools	
		Lack of investment readiness
Product Design	Sustainability-linked loans	
		Guarantees
		Insurance
Policy Actions	Develop a sustainable finance national taxonomy	
	Create and nurture a better data environment	
		Promote private participation through financial incentives

Quadrant 4: QUESTION MARKS

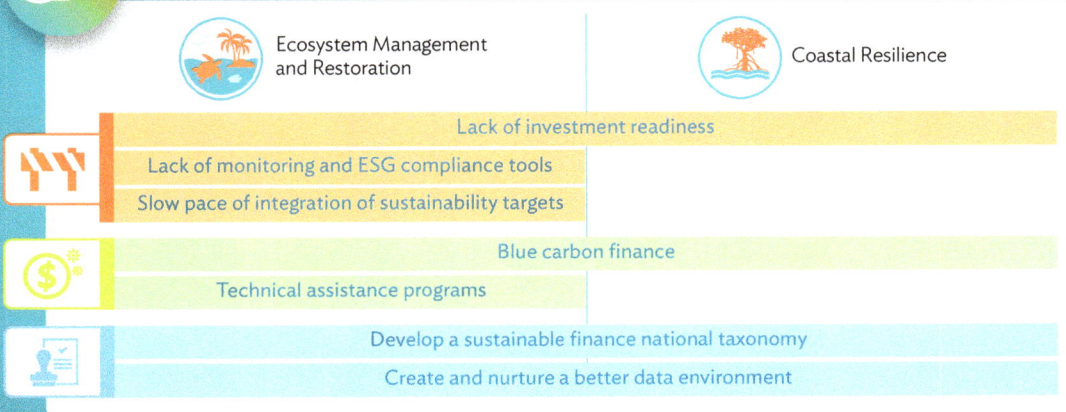

	Ecosystem Management and Restoration	Coastal Resilience
Barriers	Lack of investment readiness	
	Lack of monitoring and ESG compliance tools	
	Slow pace of integration of sustainability targets	
Product Design	Blue carbon finance	
	Technical assistance programs	
Policy Actions	Develop a sustainable finance national taxonomy	
	Create and nurture a better data environment	

ESG = environmental, social, and governance; SMEs = small and medium-sized enterprises.
Source: ADB based on AWR Lloyd.

Fishing boats lie docked at
the Phuket Fishport. Thailand
is the third-largest exporter
of seafood in the world.

I. OVERVIEW OF THE BLUE ECONOMY SECTORS IN THAILAND

Thailand's ocean economy contributes approximately 30% to its gross domestic product (GDP), the second highest in the Indo-Pacific region.[1] The country's 23 coastal provinces support the livelihoods of 24% of the total population and its major industries rely on ocean sectors such as ports and shipping. Given the importance of the ocean economy to Thailand, ensuring the sustainable use of the ocean's natural resources is necessary for safeguarding future economic growth and achieving developed country status by 2037.[2] Encouragingly, Thailand is undertaking initiatives to transition to a blue economy. For instance, in May 2022, the Department of Marine and Coastal Resources (DMCR) published the Government of Thailand's first blue economy framework, which progressively sets out activities for the Trat province including rehabilitating marine and coastal resources, strengthening knowledge management and capacity building efforts, and improving livelihoods.[3]

Table: Country Overview

Population (2019)	69.95 million
Population in Coastal Provinces	24%
Total Land Area	511,000 km
Total Maritime Zone	323,000 km
Coastline	Total ~ 3,200 km Gulf of Thailand ~ 2,100 km Andaman Sea ~1,100 km
Coastal Provinces	23 coastal provinces out of 77 provinces in total
Provinces with Major Coastal Activities	11 provinces: Trat, Rayong, Chonburi, Bangkok, Ranong, Phang Nga, Phuket, Surat Thani, Krabi, Nakhon Si Thammarat, Trang.
GDP (2021)	$506 billion
GDP per capita (2021)	$7,233
Value of Ocean Economy (2021)	$150 billion
Share of Ocean Economy to GDP	30%

GDP = gross domestic product, km = kilometer.
Note: The value of ocean economy for 2021 is extrapolated from the 2015 share of ocean economy to GDP % as stated in Partnerships in Environmental Management for the Seas of East Asia, Department of Marine and Coastal Resources. 2019. *National State of Oceans and Coasts 2018: Blue Economy Growth of Thailand.* https://seaknowledgebank.net/sites/default/files/NSOC%20Thailand%202018%20%28FINAL%29%20 12032020.pdf.
Source: AWR Lloyd.

1   Asian Development Bank. 2021. *Financing the Ocean Back to Health in Southeast Asia.* Manila. https://www.adb.org/sites/default/files/publication/756686/financing-ocean-health-southeast-asia.pdf.

2   Government of Thailand, Office of the National Economic and Social Development Board, National Strategy Secretariat Office. 2018. *National Strategy 2018–2037 (Summary).* https://www.bic.moe.go.th/images/stories/pdf/National_Strategy_Summary.pdf.

3   Government of Thailand, Ministry of Natural Resources and Environment, Department of Marine and Coastal Resources. 2022. *Guidelines for Utilization of Marine and Coastal Resources under the Blue Economy Framework in Trat Province.* https://www.dmcr.go.th/detailLib/6384.

This investment report classifies the blue economy into the sectors and subsectors described in Figure 1, based on the Ocean Finance Framework of the Asian Development Bank (ADB), and illustrates their current and projected market size in Figure 2.[4]

Figure 1: ADB Blue Economy Sectors and Subsectors

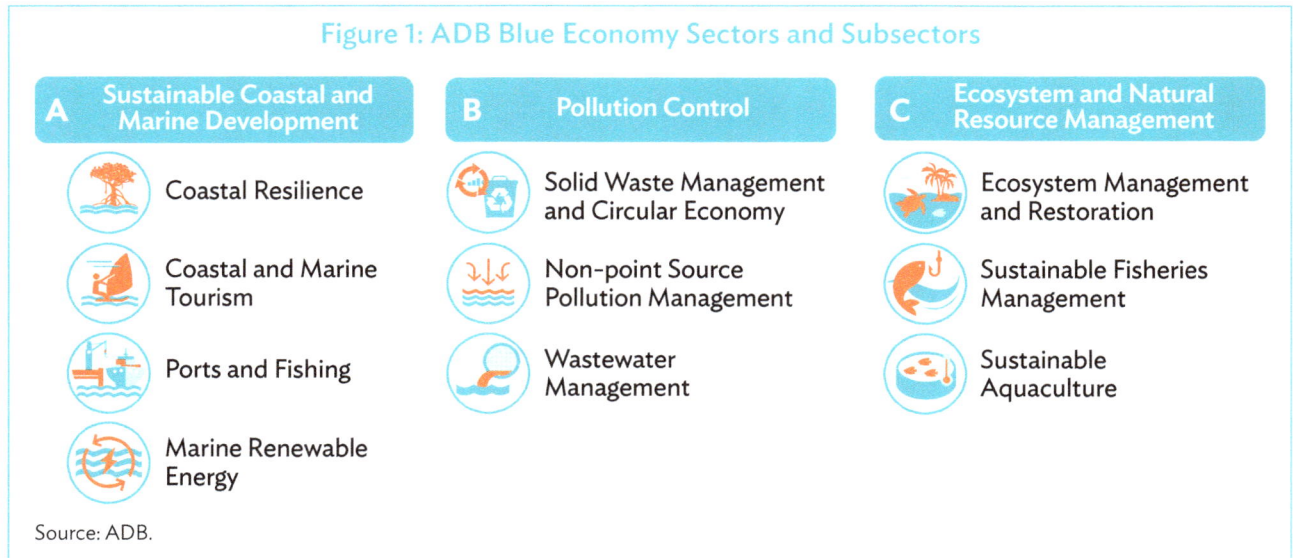

A Sustainable Coastal and Marine Development
- Coastal Resilience
- Coastal and Marine Tourism
- Ports and Fishing
- Marine Renewable Energy

B Pollution Control
- Solid Waste Management and Circular Economy
- Non-point Source Pollution Management
- Wastewater Management

C Ecosystem and Natural Resource Management
- Ecosystem Management and Restoration
- Sustainable Fisheries Management
- Sustainable Aquaculture

Source: ADB.

Figure 2: Market Size Estimate of the Blue Economy in Thailand, by Sector (billion Thai baht)

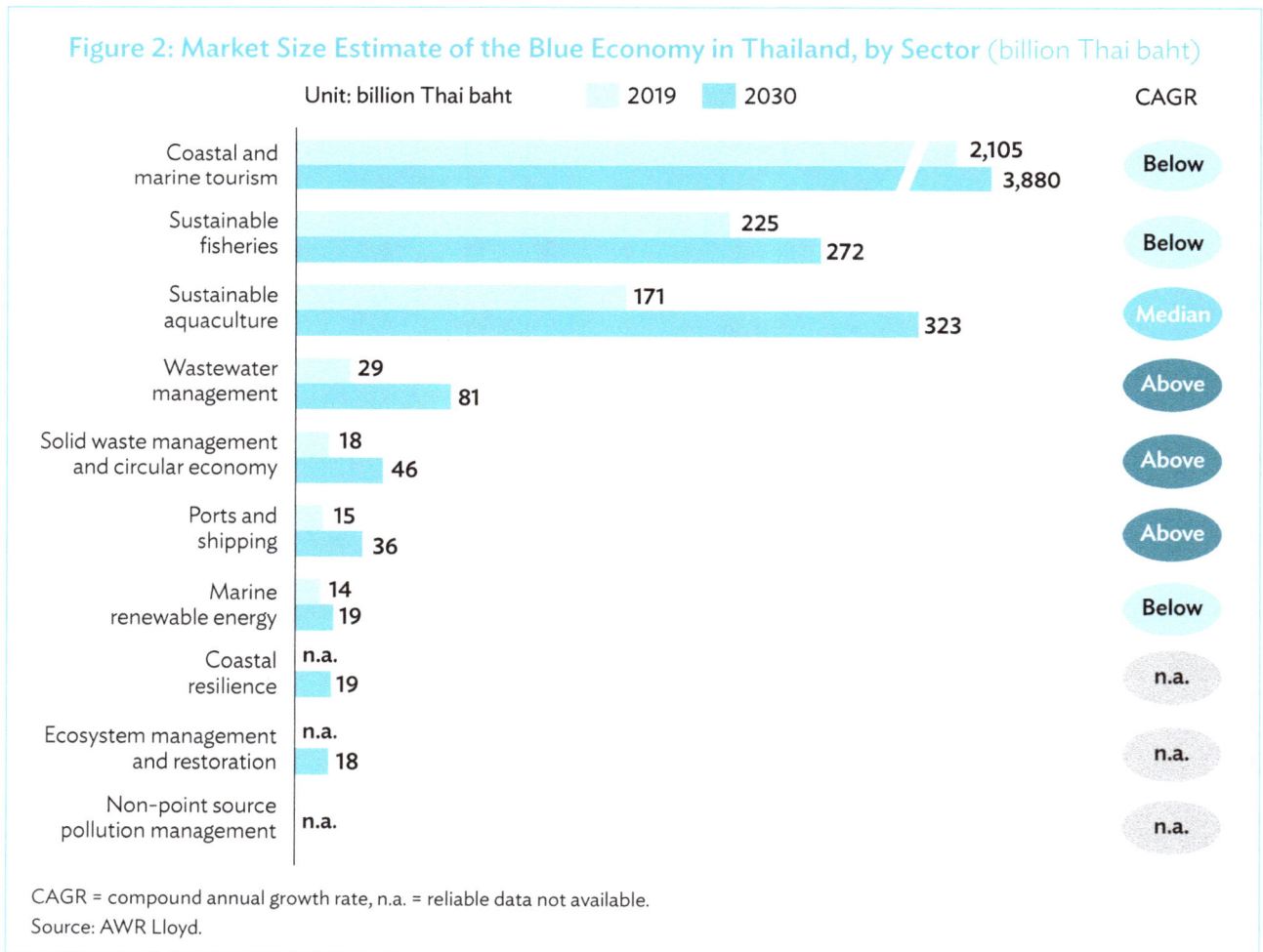

Unit: billion Thai baht   ☐ 2019   ☐ 2030

Sector	2019	2030	CAGR
Coastal and marine tourism	2,105	3,880	Below
Sustainable fisheries	225	272	Below
Sustainable aquaculture	171	323	Median
Wastewater management	29	81	Above
Solid waste management and circular economy	18	46	Above
Ports and shipping	15	36	Above
Marine renewable energy	14	19	Below
Coastal resilience	n.a.	19	n.a.
Ecosystem management and restoration	n.a.	18	n.a.
Non-point source pollution management	n.a.		n.a.

CAGR = compound annual growth rate, n.a. = reliable data not available.
Source: AWR Lloyd.

---

4   ADB. 2022b. *Asian Development Bank Ocean Finance Framework*. Manila.

The following section discusses the socioeconomic significance, key activities, and opportunities and challenges related to each subsector of Thailand's blue economy.

Sustainable Coastal and Marine Development

Within this sector, coastal and marine tourism is the largest subsector, which is projected to reach a market size of B3.9 trillion ($111 billion) by 2030, as shown in Figure 2. Ports and shipping is the second largest, with a projected market size of B36 billion ($1.02 billion) by 2030, and primarily focuses on strengthening maritime decarbonization strategies and sustainable best practices in operations. The coastal resilience subsector is comprised of noncommercial projects at the provincial level and requires significant public sector support. Marine renewable energy is the smallest subsector with growth limited due to a range of technical challenges that constrain project development.

The last section of Route 3 will be built across the Mekong River, between Houey Xai in Lao PDR and Chiang Khong in Thailand within the next few years.

Coastal Resilience

Coastal resilience cuts across several important industries critical to other blue economy sectors. Coastal erosion causes Thailand to lose approximately 30 square kilometers of land annually.[5] Meanwhile, it is estimated that coastal flooding could affect more than 2.4 million people between 2070 and 2100.[6] More immediately, Greenpeace estimates that if Bangkok sinks below sea level due to its vulnerability to flooding, Thailand would lose up to $500 billion in GDP.[7] The baseline review determined that both traditional and ecosystem-based adaptation efforts are being implemented to reduce coastal erosion and build social resilience in coastal communities.

The coastal resilience sector relies heavily on public sector participation with mostly noncommercial, provincial projects. All the projects identified in the baseline review were developed by the public sector and primarily focus on mitigation. Funding for mitigation is applied toward integrated coastal management, which incorporates enhanced climate risk planning and the reduction of pollution that leads to the degradation of marine resources. In addition, the funds are being used to strengthen community resilience, which includes the prevention of destructive fishing practices and pathways for fishing communities to conserve and manage coastal ecosystems and enhance local resilience. Research and policy work, including the development of a National Adaptation Plan to address adaptation measures for Thailand's marine and coastal areas, also received government support.[8]

The DMCR allocated more than B6 billion ($171 million) for over 70 projects in 2013–2020.[9] Projects operationalized in this sector, such as seawall and revetment construction initiatives, are nonrevenue generating and thus unattractive to private investors. The sector has room for innovation and growth, and any efficiencies arising thereof, may raise interest from the private sector, which has previously financed related sectors such as tourism and ports and shipping.

[5]   Mahidol University, Faculty of Environment and Resource Stuides. 2022. *Thai Coast: Vulnerability, Resilience, and Adaptation in Thailand.* https://en.mahidol.ac.th/index.php/sdgs/goal-13-climate-action/94-sdg/2565/2022-10-04-06-16-55/746-thai-coast-coastal-vulnerability-resilience-and-adaptation-in-thailand.

[6]   World Bank and Asian Development Bank. 2021. *Thailand Climate Risk Country Profile.* https://climateknowledgeportal.worldbank.org/sites/default/files/2021-08/15853-WB_Thailand%20Country%20Profile-WEB_0.pdf.

[7]   Greenpeace. 2021. *The Projected Economic Impact of Extreme Sea-Level Rise in Seven Asian Cities in 2030.* https://www.greenpeace.org/static/planet4-eastasia-stateless/2021/06/966e1865-gpea-asian-cites-sea-level-rise-report-200621-f-3.pdf.

[8]   Green Climate Fund. 2020. *Increasing Resilience to Climate Change Impacts in Marine and Coastal Areas along the Gulf of Thailand.* https://www.greenclimate.fund/sites/default/files/document/tha-rs-006.pdf.

[9]   A. Wipatayotin. 2021. New Coastal Checklist Gets the Nod. *Bangkok Post.* 28 September. https://www.bangkokpost.com/thailand/general/2188571/new-coastal-checklist-gets-the-nod.

Coastal and Marine Tourism

Before the coronavirus disease (COVID-19) pandemic, Thailand's tourism industry accounted for 11%–20% of the country's GDP, contributed over $10 billion in revenue, and supported around 20% of total employment.[10] These indicators are expected to improve post-pandemic due to the completion of various infrastructure projects that enable greater connectivity across the country. One such example is the Eastern Economic Corridor (EEC) high speed rail, which links Don Mueang, Suvarnabhumi, and U-Tapao international airports.[11]

By virtue of its scale, the environmental impacts of activities in this sector such as marine pollution and damage to ecosystems and habitats are significant. As of 2018, over 70% of coral reefs in the country have been lost.[12] Moreover, the over-exploitation of marine resources by the tourism industry and local fisheries is being exacerbated by the increasing extraction of reef fish.[13] To address these challenges, Thailand is increasingly pivoting toward sustainable tourism aimed at integrating sustainable best practices, including adopting standards recognized by the Global Sustainable Tourism Council.[14]

Coastal and marine tourism is particularly sensitive to climate change, and adaptation and mitigation efforts are equally relevant. Most of the projects identified are operational and are located across the coastal provinces near the Andaman Sea, such as Krabi, Phang Nga, and Phuket. While these projects have been primarily publicly financed, resorts and hotels are also independently launching and financing initiatives. Projects aimed at mitigation measures focus on: (i) de-prioritizing traditional fuel sources, (ii) developing a surveillance system for marine pollution, (iii) improving waste management practices, and (iv) conserving or restoring marine wildlife. Some adaptation measures in the sector focus on the development of recreational activities, such as mangrove learning centers, that strengthen research and education and boost sustainable tourism. Protecting marine wildlife and expanding health- and wellness-related tourism are also addressed through adaption projects.

Ports and Shipping

In 2021, the Port Authority of Thailand (PAT) reported revenues of more than $440 million generated from the ports and shipping sector, approximately 70% of which came from shipping.[15] Due to its lower cost base and operational efficiencies compared to land and air transportation, maritime transportation accounts for approximately 80% of the county's international trade.[16] In 2020, approximately 1.33 million people were employed in transportation and storage sectors.[17] The three major ports in Thailand are Bangkok Port, Laem Chabang Port, and Map Ta Phut Industrial Port. Laem Chabang Port alone represents over 50% of the total port capacity of Thailand.[18] In 2020, Laem Chabang Port ranked 20th among the world's largest container ports with a total annual volume of 7.55 million twenty-foot

[10] Statista. 2022b. Tourism Industry in Thailand – Statistics & Facts, International and Domestic Tourists' Revenue. https://www.statista.com/topics/6845/tourism-industry-in-thailand/#topicHeader___wrapper (accessed February 2023).

[11] *Bangkok Post*. 2023. 3-Airport High-speed Rail Link Completion Seen by 2029. 29 January. https://www.bangkokpost.com/thailand/general/2493950/3-airport-high-speed-rail-link-completion-seen-by-2029.

[12] *Deutsche Welle*. 2018. Thailand's Tourism Boom Damages Corals. 30 January. https://www.dw.com/en/thailands-tourism-boom-damages-corals-to-critical-level/a-42361812.

[13] Partnerships in Environmental Management for the Seas of East Asia, Department of Marine and Coastal Resources. 2019. *National State of Oceans and Coasts 2018: Blue Economy Growth of Thailand*. https://seaknowledgebank.net/sites/default/files/NSOC%20Thailand%202018%20%28FINAL%29%2012032020.pdf.

[14] The Board of Investment of Thailand. 2020. *Opportunities in the Bio-Circular-Green (BCG) Economy and BOI Support Measures*. https://www.boi.go.th/upload/content/BOI-BCG_DSG%20Chokedee.pdf; TTG Asia. 2018. Thai Sustainable Tourism Standard Gets GSTC Recognition. 11 May. https://www.ttgasia.com/2018/05/11/thai-sustainable-tourism-standard-gets-gstc-recognition/221195/.

[15] *The Nation Thailand*. 2022. Port Authority Hoists Revenue to Bt15.6 Billion in 2021. 4 March. https://www.nationthailand.com/business/40013020.

[16] Kingdom of Thailand. 2021. *Thailand's Overall Maritime Strategy (unofficial translation)*. https://md.go.th/wp-content/uploads/2021/03/Thailands-Overall-Maritime-Strategy.pdf; J. Rudjanakanoknad, W. Suksirivoraboot, and S. Sukdanont. 2014. Evaluation of International Ports in Thailand through Trade Facilitation Indices from Freight Forwarders. *Procedia – Social and Behavioral Sciences*. 111 (2014). pp. 1073–1082. https://www.sciencedirect.com/science/article/pii/S1877042814001438.

[17] Government of Thailand, Office of the National Economic and Social Development Council. 2020. *Thailand's Logistics Report 2020*. https://www.nesdc.go.th/ewt_dl_link.php?nid=11975.

[18] The Board of Investment of Thailand. 2021. *Seaports*. https://www.boi.go.th/index.php?page=seaports.

equivalent units (TEUs).[19] With port expansion plans in the pipeline, adopting sustainability strategies that address environmental concerns such as oil discharge, ship waste, and carbon emissions are vital.

The projects identified in the baseline review are mostly public sector-led, large scale, and focused on mitigation. Improving the sustainability of port operations by transitioning away from traditional fuel sources toward more efficient technologies and adopting waste management best practices are central to this subsector. In addition, ecosystem restoration activities that preserve and rehabilitate 4.5 hectares of mangrove forests situated within the Laem Chabang Port area were also identified.[20]

Marine Renewable Energy

Thailand's geographical characteristics are well-suited for traditional solar and wind projects. Additionally, marine renewable energy (generated from the natural movement of water through waves, tides, and river and ocean currents, and includes offshore wind and floating solar panels deployed at sea) can be developed as a power source. By 2037, Thailand's power demand is expected to be over 56,000 megawatts (MW).[21] The contribution of energy from marine renewables toward this need remains insignificant with just over 2,700 MW anticipated to be supplied by solar and hydropower and 1,400 MW from wind energy.[22] It is anticipated that a shift to 100% renewable energy could create over 170,000 jobs in the country by 2050 (including up to 10,000 jobs from offshore and onshore wind power projects).[23]

The Government's Thailand 4.0 and Energy 4.0 strategies aim to transform Thailand into a low-carbon country based on clean and alternative energy, prioritizing energy efficiency and storage, and promoting the use of electric vehicles. As of 2021, Thailand's installed renewable energy capacity of approximately 15 gigawatts (GW) accounts for one-third of the overall power mix. By 2030, this capacity is forecast to grow to 63 GW and a 39% share, which would position the country as one of Southeast Asia's regional leaders for renewable energy.[24] Furthermore, a framework on energy efficiency, funding for research and development (R&D) in energy storage, and an enabling environment for distributed power generation are being put in place to achieve the country's long-term goals on renewable energy.[25]

The marine renewables sector has not been included, however, in the nation's Power Development Plan. This is likely because investment costs and risks for private investors remain high due to the sector's early stage and fragmented nature. Moreover, the country's geography and resulting variability of wind speed increases overall costs related to offshore wind generation.[26] Hence, such projects require public and concessional funding to reduce the risk profile and strengthen commercial viability.

Initiatives in this sector have focused equally on transition and R&D. It is notable that the Electricity Generating Authority of Thailand (EGAT) is supporting the development of floating hydro solar farms, which combine marine and renewable energy technology, on dams (not seas or oceans) across Thailand.[27] Development of and access to offshore grid and distribution systems are needed to harness marine renewable energy.

[19] World Shipping Council. *The Top 50 Container Ports*. 2020. https://www.worldshipping.org/top-50-ports.

[20] Partnerships in Environmental Management for the Seas of East Asia. 2019. *ICM Solutions, Gateway to a Blue Economy: Port Safety, Health, and Environmental Management in the Port Authority of Thailand— Bangkok and Laem Chabang Ports.* http://pemsea.org/sites/default/files/KP%2024_0_0.pdf.

[21] The Board of Investment of Thailand. 2020b. *Thailand's Electrical Market.* https://www.boi.go.th/index.php?page=electricity.

[22] Government of Thailand, Energy Policy and Planning Office. 2018. *Thailand's Power Development Plan (PDP) 2018-2037 (TH).* https://policy.thinkbluedata.com/sites/default/files/Thailand%E2%80%99s%20Power%20Development%20Plan%20%28PDP%29%20%282018%E2%80%932037%29%20%28TH%29.pdf.

[23] Greenpeace. 2018. *Renewable Energy Job Creation in Thailand.* https://www.greenpeace.or.th/report/Renewable-Energy-Job-Creation-in-Thailand-EN.pdf.

[24] A. Modi and M. Lackovic. 2021. Investment and Innovation in Thai Renewable Energy. *Bangkok Post.* 15 March. https://www.bangkokpost.com/business/2083795/investment-and-innovation-in-thai-renewable-energy.

[25] Asian Development Blog. 2019. *Thailand in Need of 'Energy 4.0.'* https://blogs.adb.org/blog/thailand-need-energy-40.

[26] M. Ranthodsang et al. 2020. Offshore Wind Power Assessment on the Western Coast of Thailand. *Energy Reports.* 6. pp. 1135–1146.

[27] EGAT. 2021. The World's Largest Hydro-floating Solar Hybrid. 1 July. https://www.egat.co.th/home/en/the-worlds-largest-hydro-floating-solar-hybrid/.

Pollution Control

Based on ADB's taxonomy, the pollution control sector is classified into three subsectors: solid waste management and circular economy, non-point source pollution management, and wastewater management. Solid waste management cuts across other sectors involved in the blue economy and hence provides attractive financing opportunities. Thus, commercial banks are relatively more invested in pollution control (though blue bonds and loans) than in the sustainable coastal development and ecosystem and natural resources management sectors.

Discarded plastic water bottles floating near a wharf. Plastics that wash in the oceans kill an estimated 1.1 million marine creatures annually.

Solid Waste Management and Circular Economy

Solid waste is a significant contributor to marine debris, detrimentally affecting marine life and ecosystems. In June 2018, in the southern province of Songkhla, 8 kilograms (kg) of plastic bags were found in the stomach of a beached pilot whale, which led to its death.[28] In 2016, solid waste disposal in Thailand accounted for over 48% of the total greenhouse gas (GHG) emissions generated by the waste sector.[29] Thailand relies heavily on single-use plastics, and consumption of such plastics only intensified during the COVID-19 pandemic. In 2020, Thailand generated 1.1 kg of waste per capita/per day resulting in over 25 million tons of municipal solid waste.[30] Approximately 2 million tons of plastic waste in the same year resulted in Thailand ranking 5th among the world's biggest marine plastic polluters.[31] Uncollected plastic waste and unsanitary disposal facilities have resulted in 428,000 tons per year of mismanaged plastic waste.[32] The insufficient sorting at source, especially among domestic households and high-density areas, and low waste source data, specifically regarding waste collection reporting at disposal and recycling facilities, are important considerations in this sector.[33]

Incorporating waste management into a circular economy model refers to the reduction, recycling, or conversion of solid waste into sustainable products to minimize environmental, social, and economic damages. The Thailand Country Private Sector Diagnostic (CPSD) report estimates that Thailand could annually generate as much as $1.6 billion in cost savings and additional revenue for the private sector by transitioning into a circular economy.[34] Stronger measures to promote the reduction of waste generation and the tracking and sorting of waste will contribute to such a transition.

The solid waste management sector has been bolstered by several public sector initiatives, such as the Bio–Circular–Green (BCG) Economy strategic plan of the Thailand Board of Investment (BOI Thailand) that supports waste management, waste–to–energy infrastructure, and other circular economy models. Innovative financing instruments are increasingly being used to fund projects in this sector. For example, the International Finance Corporation (IFC), ADB, and other development finance institutions have issued a blue loan to Indorama Ventures, supporting the company's recycling of 50 billion polyethylene terephthalate (PET) bottles annually by 2025 in its operating countries including Thailand. This project assigns economic value to waste, thereby incentivizing waste collection systems and ultimately diverting plastic waste from the marine environment.

---

[28] D. Marks, M. A. Miller, and S. Vassanadumrongdee. 2020. The Geopolitical Economy of Thailand's Marine Plastic Pollution Crisis. Asia Pacific Viewpoint. 61 (2). pp. 266–282. https://onlinelibrary.wiley.com/doi/full/10.1111/apv.12255.

[29] United Nations Framework Convention on Climate Change. 2021. Mid-century, Long-term Low Greenhouse Gas Emissions Development Strategy, Thailand. https://unfccc.int/sites/default/files/resource/Thailand_LTS1.pdf.

[30] Thailand Environment Institute. 2021. Waste to Energy. https://www.tei.or.th/en/article_detail.php?bid=49.

[31] P. Tanakasempipat. 2020. Plastic Piles Up in Thailand as Pandemic Efforts Sideline Pollution Fight. Reuters. 11 May. https://www.reuters.com/article/us-health-coronavirus-thailand-plastic-idUSKBN22N12W.

[32] World Bank. 2022. Plastic Waste Material Flow Analysis for Thailand. https://documents1.worldbank.org/curated/en/099515103152238081/pdf/P17099409744b50fc09e7208a58cb52ae8a.pdf.

[33] Thailand Development Research Institute. 2019. Tackling Thailand's Food-waste Crisis. https://tdri.or.th/en/2019/10/tackling-thailands-food-waste-crisis/.

[34] S. Toomgum. 2022. Thais Urged to Embrace Disruptive Tech. Bangkok Post. 23 February. https://www.bangkokpost.com/business/2268483/thais-urged-to-embrace-disruptive-tech.

Non-point Source Pollution Management

Non-point source (NPS) pollutants include toxins, pathogens, excess nutrients, and sediments, that run off from farmlands, city streets, roofs, and construction sites, and subsequently enter natural waterways as invasive species. For example, industrial growth in Thailand has resulted in the continual decline of groundwater quality due to toxic chemical discharge from factories, oil leaks, and submerged cargo ships.[35]

The major sustainability issues in the sector include: (i) severe soil erosion that is prevalent in approximately 33% of the total geographical land; (ii) habitat alteration that impacts local ecosystems and species, water temperatures, and sediment erosion; and (iii) data inaccuracy due to limited tools to measure NPS pollutants.[36]

While there is no reliable disclosure of public expenditure specifically for NPS pollution management, most projects focus on pollution management and the development of a data collection system for pollution tracking, especially for the public sector.[37] NPS pollution management is addressed mainly through activities under the wastewater management sector. Green infrastructure developments that help filter NPS runoffs is an example.

Wastewater Management

Wastewater management refers to the treatment of wastewater to remove any contaminants that could harm the environment and public health, while ensuring economic, social, and political soundness. It also includes initiatives and practices to collect, recycle, and reuse treated water to conserve natural resources of clean water. In 2016, approximately 15% of the total volume of water consumed in Thailand came from wastewater treatment systems.[38]

As a result of increasing investments from public and private stakeholders, an estimated 10% annual growth in wastewater reuse is anticipated (footnote 38). For the 2019–2023 period, the National Budget for Environmental Project in Wastewater Management has allocated up to $2.6 billion toward wastewater treatment plans in 93 communities.[39]

The key sustainability issues in the sector include: (i) poor surface water quality, where around 14% of surface water sources are considered poor due to improper water drainage in households, tourist attractions, and other areas;[40] (ii) excessive groundwater extraction causing the intrusion of seawater inland to increase salinity levels in the remaining freshwater; and (iii) lack of awareness regarding wastewater dumping (e.g., food stalls each drain, on average, 4 liters of wastewater comprised of oil and other residues daily into sewage systems, canals, and rivers).[41] Additionally, the country's municipal wastewater contributes to around 47.5% of the total GHG emissions from the waste sector.[42]

[35] J. A. U. Frias and R. Kumar. 2022. Creating Markets in Thailand: Rebooting Productivity for Resilient Growth. Country Private Sector Diagnostic (CPSD) Washington, DC: World Bank Group. https://documents.worldbank.org/en/publication/documents-reports/documentdetail/468721645451588595/creating-markets-in-thailand-rebooting-productivity-for-resilient-growth.

[36] S. Semmahasak. 2014. Soil Erosion and Sediment Yield in Tropical Mountainous Watershed of Northwest Thailand: The Spatial Risk Assessments Under Land Use and Rainfall Changes. School of Geography, Earth and Environmental Sciences, College of Life and Environmental Sciences, University of Birmingham. https://core.ac.uk/reader/33527420; NBSAP Forum. 2018. Combating Industrial Pollution in Thailand. https://nbsapforum.net/knowledge-base/best-practice/combating-industrial-pollution-thailand.

[37] Enviliance Asia. 2022. Another Step Toward PRTR Implementation in Thailand. https://enviliance.com/regions/southeast-asia/th/th-chemical/th-prtr.

[38] Kingdom of the Netherlands, Open Development Mekong. 2016. The Water Sector in Thailand. https://data.opendevelopmentmekong.net/dataset/379173fa-bc66-4866-87ca-00cc73e8139f/resource/b9547f82-089d-438c-91e8-8cc87d860dc4/download/factsheet-the-water-sector-in-thailand-3.pdf.

[39] L. Apisitniran. 2019. Treatment Touted for 10% Growth. Bangkok Post. 11 March. https://www.bangkokpost.com/business/1642368/treatment-touted-for-10-growth.

[40] Statista. 2022a. Water Quality of Surface Water Surfaces in Thailand 2021. https://www.statista.com/statistics/1295279/thailand-condition-of-surface-water-sources/.

[41] S. Sukphisit. 2019. Troubled Waters. Bangkok Post. 28 July. https://www.bangkokpost.com/life/social-and-lifestyle/1720327/troubled-waters.

[42] Thai–German Cooperation. 2018. ECAM Tool Gears Up Water Sector Towards Achieving Greenhouse Gas Reduction Target. 20 April. https://www.thai-german-cooperation.info/en_US/ecam-tool-gears-up-water-sector-towards-achieving-greenhouse-gas-reduction-target/.

Wastewater management is one of the more mature and commercial sectors in Thailand's blue economy. The private sector has used innovative financing instruments (such as green bonds, green loans, and sustainability-linked loans) to support resource conservation projects, procurement of wastewater treatment technologies, and other recycling activities. Blue finance mechanisms, such as the maritime sustainability bonds issued by TMB Thanachart Bank and subscribed by IFC, can be structured to finance water conservation and wastewater treatment projects.[43]

Ecosystem and Natural Resources Management

According to ADB's taxonomy, ecosystem and natural resources management is comprised of three subsectors: ecosystem management and restoration, sustainable fisheries, and sustainable aquaculture. Public sector initiatives in this area have focused on decarbonization. As a result of increased global demand for seafood, private sector investments have been directed primarily toward sustainable fisheries and sustainable aquaculture.

Ecosystem Management and Restoration

The preservation and restoration of natural ecosystems and biodiversity from further degradation caused by climate change and human activities is central to this sector. Several of Thailand's key industries including tourism, fisheries, and aquaculture are dependent on healthy marine ecosystems and natural capital. There is, however, a gap in biodiversity financing of around $1.16 billion to achieve the country's National Biodiversity Strategic and Action Plan.[44] Thailand's mangrove ecosystems are also critical in alleviating the effects of climate change but, since 1961, more than 50% of its mangrove forests have been lost due to the expansion of shrimp and salt farms.[45] Green

A fisher shows off his catch at Phuket's Rawai Beach. Fishing is a way of life for many locals in Southern Thailand.

growth policies instituted by the government are estimated to contribute almost $2 billion toward Thailand's national economy, with ecosystem services and capital supplied by forests, wetlands, mangroves, and coral reefs.[46] Thailand has also set a national target to protect up to 10% of coastal and marine areas under Marine Protected Areas (MPAs).[47]

The main sustainability issues in this sector include: (i) threat to important ecosystems and species, (ii) pollution of aquatic environments, and (iii) aquatic resource and habitat depletion. Rapid coastal development and the unsustainable use of resources have continuously threatened mangrove, coral reef, and seagrass ecosystems, and endangered species. Additionally, multiple sources of pollutants, such as sewage-borne pathogens, oil pollution, marine debris, and toxic substances, are also dispersed in the marine environment (footnote 13). These factors, along with overcapitalization and overexploitation by coastal communities, have resulted in the continued degradation of marine resources and habitats.

---

43   Hunton Andrews Kurth. 2022. Hunton Andrews Kurth Secures First Maritime Sustainability Bond in Thailand for TMB Thanachart. 2 June. https://www.huntonak.com/en/news/hunton-secures-first-maritime-sustainability-bond-in-thailand-for-tmb-thanachart-with-intl-finance-corporation-as-the-subscriber.html.

44   The Biodiversity Finance Initiative. 2018. Thailand. https://www.biofin.org/thailand.

45   E. B. Barbier. 2006. Mangrove Dependency and the Livelihoods of Coastal Communities in Thailand. In C. T. Hoanh, T. P. Tuong, J. W. Gowing, and B. Hardy, eds. *Environment and Livelihoods in Tropical Coastal Zones: Managing Agriculture-Fishery-Aquaculture Conflicts*. http://dx.doi.org/10.1079/9781845931070.0126.

46   *Open Development Mekong, Thailand*. 2017. Environment and Natural Resources. 19 December. https://thailand.opendevelopmentmekong.net/topics/environment-and-natural-resources/.

47   United Nations Department of Economic and Social Affairs. 2016. *Thailand towards Sustainable Management of Marine and Coastal Habitats*. https://sdgs.un.org/partnerships/thailand-towards-sustainable-management-marine-and-coastal-habitats.

Public–private partnerships have spearheaded mangrove reforestation and coral restoration projects. While commercial opportunities are relatively untapped, the DMCR has begun developing a 10-year, 300,000 rai mangrove plantation project. This initiative is expected to enable higher private sector participation in planting mangroves for carbon credits and to contribute toward Thailand's target of reaching Net Zero by 2065.[48]

Sustainable Fisheries Management

The sustainable fisheries management sector endeavors to maintain the natural balance of marine ecosystems by harvesting fish at rates that are appropriately capped by the responsible authorities. Thailand has been a leading marine fisheries producer and exporter for several decades with fishery products comprising about 20% of the country's total food product exports.[49] Local consumption is also high as fish is the primary source of animal protein for most of the population, especially in the coastal provinces.[50] In 2020, Thailand's Cabinet approved the Sea Fisheries Management plan for 2020–2022, with a budget of over B2.9 billion ($82.6 million) to continue efforts initiated in its 2015 plan to support Thai fisheries.[51]

The sustainability considerations in the sector include: (i) overfishing and illegal, unreported, and unregulated fishing (IUU), which directly affects the abundance of bycatch species and depletes fishing stocks; (ii) high levels of marine debris such as fishing gear and microplastics, which have disrupted marine life and threatened endangered species; and (iii) lack of appropriate fishery subsidies. Although Thailand ranked 7th in the world for fisheries subsidies totaling B36 billion ($1.02 billion), only around 6% of the subsidies budget in 2018 was allocated for conservation with the majority allocated to capacity-enhancing activities such loans to help build boats and fuel subsidies. These subsidies can lead to overfishing and the further degradation of the marine environment.[52] Thailand has a significant opportunity to incentivize activity in support of the blue economy by redirecting these subsidies. Greater engagement in community awareness and improved training are also required to improve resiliency among local communities and ecosystems.

Significant private sector traction contributes to the commercial value of sustainable fisheries management, especially with the adoption of blue financing products by Thai Union, one of the largest seafood producers in the world. As of 2021, the company has pioneered the development of blue finance in Thailand through its B27 billion ($769.2 million) sustainability-linked loan and sustainability-linked bond.[53]

---

[48]   *Thansettakij.* 2022. Mangrove Planting to Reduce Carbon Credits. 11 May. https://www.thansettakij.com/business/524532.

[49]   Partnerships in Environmental Management for the Seas of East Asia, Department of Marine and Coastal Resources. 2019. *National State of Oceans and Coasts 2018: Blue Economy Growth Thailand.* http://pemsea.org/sites/default/files/NSOC%20Thailand%202018%20(FINAL)%20 12032020.pdf; Government of the United States, Department of Agriculture, Foreign Agriculture Service, Global Agricultural Information Network. 2018. Thailand Seafood Report. https://apps.fas.usda.gov/newgainapi/api/report/downloadreportbyfilename?filename=Seafood%20 Report_Bangkok._Thailand_5-8-2018.pdf.

[50]   S. Suwannapoom. 2020. County Fisheries Trade: Thailand. *Southeast Asian Fisheries Development Center (SEAFDEC).* 2 April. http://www.seafdec.org/county-fisheries-trade-thailand/.

[51]   Government of Thailand, Ministry of Agriculture and Cooperatives, Department of Fisheries. 2019. *Marine Fisheries Management Plan of Thailand 2020–2022.* https://faolex.fao.org/docs/pdf/tha212512.pdf.

[52]   D. Chakrabongse. 2021. Thailand's Subsidies are Now the Biggest Threat to its Fisheries. *Thai Enquirer.* 8 April. https://www.thaienquirer.com/26235/thailands-subsidies-are-now-the-biggest-threat-to-its-fisheries/.

[53]   Thai Union. 2022. Healthy Living Healthy Oceans. *56-1 One Report.* https://investor.thaiunion.com/misc/ar/20220325-tu-or2022-en.pdf.

Sustainable Aquaculture

Sustainable aquaculture refers to the controlled cultivation of aquatic organisms for cultural practices. Due to the increasing domestic and international demand, aquaculture production (particularly marine shrimp) has significantly contributed to the socioeconomic development of Thailand.[54] According to the Southeast Asian Fisheries Development Center (SEAFDEC), aquaculture businesses directly and indirectly employed more than 650,000 people in 2017.[55] Moreover, during 2012–2022, the shrimp industry engaged more than 1 million entrepreneurs and households and generated nearly B100 billion ($2.8 billion) a year.[56]

The main sustainability issues in the sector include: (i) depletion of mangrove forests due to increasing expansion of shrimp farms and the introduction of alien species; (ii) saline contamination of soil and inland water bodies from aquaculture effluents that contain nitrogen, phosphorus, and saline; and (iii) outbreak of diseases among aquaculture species that create potential reservoirs for viruses.[57]

Thailand was the world's top exporter of shrimp, with production peaking at more than 600,000 metric tons in 2011. However, according to the Thai Shrimp Association, poor water quality and the industry's rapid growth and unsustainable practices, have led to persistent outbreaks of epidemic diseases such as acute hepatopancreatic necrosis disease, previously known as early mortality syndrome. These epidemic diseases have resulted in an estimated average loss of approximately B100 billion per year for the last decade on shrimp exports and in an increasing dependence on imports.[58] Encouragingly, by the end 2023, Thailand is expected to increase domestic production of white shrimp to 400,000 metric tons, from an estimated total production of 320,000 metric tons in 2022.[59]

Public sector activities recognize the importance of the aquaculture industry and are implementing fisheries policies that will prevent aquatic animal diseases and promote environmentally friendly practices. The Department of Fisheries' Aquatic Bank program aims to also enhance the skills of local communities through the creation of over 20 development areas to increase sustainable aquaculture production across Thailand. Private sector initiatives in this sector include market leader Thai Union's 2021 investment of approximately B250 million ($7.1 million) to expand its aquaculture production capacity.[60] The company is the first in Thailand to produce and distribute feeds in the form of pellets to replace the use of fresh prey. Charoen Pokphand Foods (CPF) has also initiated projects to ensure the responsible use of resources in its aquatic feed production.[61]

---

[54] T. Sampantamit et al. 2020. Aquaculture Production and Its Environmental Sustainability in Thailand: Challenges and Potential Solutions. *Sustainability.* 12 (5). https://doi.org/10.3390/su12052010.

[55] T. Yenpoeng. 2017. Fisheries Country Profile: Thailand. *SEAFDEC.* http://www.seafdec.org/fisheries-country-profile-thailand/.

[56] *Bangkok Post.* 2022b. Shrimp Industry Continues to Tread Water. 15 December. https://www.bangkokpost.com/business/2460805/shrimp-industry-continues-to-tread-water.

[57] A. K. H. Priyashnatha and U. Edirisinghe. 2021. Lessons Learnt from the Past to Mitigate the Negative Aspects of Aquaculture in Developing Countries. *PSAKU International Journal of Interdisciplinary Research.* 10 (2). https://papers.ssrn.com/sol3/papers.cfm?abstract_id=3964762.

[58] *The Nation Thailand.* 2022a. Recurring Disease has Taken a Heavy Toll on Thai Shrimp Yields. 16 February. https://www.nationthailand.com/blogs/business/40012418.

[59] *The Nation Thailand.* 2022d. Thailand to Import 10,500 Tonnes of Shrimp as Domestic Yield Sinks. 8 August. https://www.nationthailand.com/business/40018647.

[60] *Thai Union.* 2021b. Thai Union Feedmill Aims to Lead Aquaculture and Commercial Feed Sector, Following Listing on Stock Exchange of Thailand. 29 October. https://www.thaiunion.com/en/thai-union-cares/leadership/1475/thai-union-feedmill-aims-to-lead-aquaculture-and-commercial-animal-feed-sector-following-listing-on-stock-exchange-of-thailand.

[61] *CPF.* 2017. CPF Reduces Fishmeal in Aquatic Feed Production, to Promote Sustainable Use of Marine Resources. 21 July. https://www.cpfworldwide.com/en/media-center/1034.

II. SUPPORT FROM FINANCIAL INSTITUTIONS

Thailand's six largest commercial banks, each with assets of more than B1 trillion ($28.5 billion), maintain some level of lending exposure to Thai blue economy sectors.[62] However, the overall banking sector remains largely opportunistic in developing blue economy portfolios. In addition, some specialized financial institutions (SFIs) have expressed interest in blue economy sectors and activities in line with their mandates. For example, the Export-Import Bank of Thailand has invested in the shipping and coastal and marine tourism sectors.

As indicated in Figure 3, support from Thailand's financial institutions for the blue economy varies between sectors. The sustainable fisheries and sustainable aquaculture sectors have the most investment activity, closely followed by solid waste management and circular economy, and wastewater management. Meanwhile, other blue economy sectors are either largely untapped or not specifically targeted by Thai financial institutions. For example, while lending toward renewable energy projects has become relatively mainstream (especially for solar photovoltaic), lending toward projects in the marine renewable energy sector is virtually nonexistent.

Figure 3: Heatmap—Support from Financial Institutions for Thailand's Blue Economy

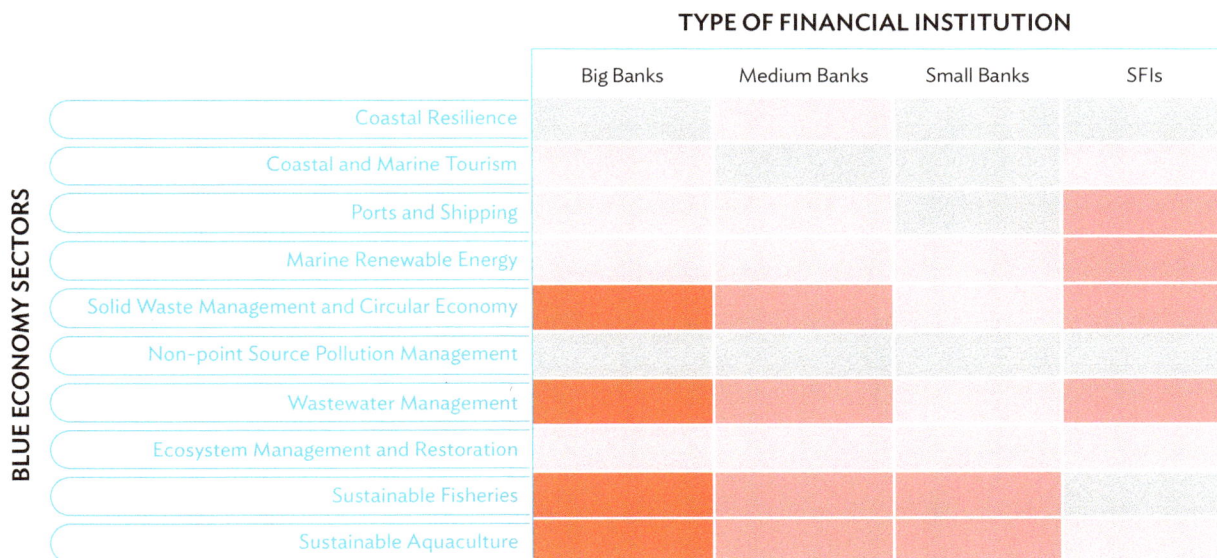

TYPE OF FINANCIAL INSTITUTION

BLUE ECONOMY SECTORS	Big Banks	Medium Banks	Small Banks	SFIs
Coastal Resilience				
Coastal and Marine Tourism				
Ports and Shipping				■
Marine Renewable Energy				■
Solid Waste Management and Circular Economy	■	■	■	■
Non-point Source Pollution Management				
Wastewater Management	■	■		■
Ecosystem Management and Restoration				
Sustainable Fisheries	■	■	■	
Sustainable Aquaculture	■	■	■	

SFI = specialized financial institution.

Notes:
1. Dark red indicates higher support from financial institutions to each subsector in the blue economy in Thailand. Grey indicates a lack of reliable data to accurately assess interest levels.
2. There are 29 commercial banks in Thailand and seven state-owned specialized financial institutions (SFIs). Asset sizes for commercial banks as of July 2022 have the following categorization: (i) big banks: assets more than B1 trillion ($28.5 billion); (ii) medium banks: assets between B200 billion ($5.7 billion) and B1 trillion ($28.5 billion); and (iii) small banks: assets less than B200 billion ($5.7 billion). Asset sizes for SFIs is sourced from latest annual reports.

Source: AWR Lloyd.

---

[62] These include Bangkok Bank, Kasikornbank, Krungthai Bank, Siam Commercial Bank, Bank of Ayudhya, and TMB Thanachart Bank.

III. SUPPORT FROM THE GOVERNMENT OF THAILAND

Despite being a coastal nation with significant socioeconomic dependence on ocean-related industries, Thailand has yet to widely adopt a robust blue economy regulatory framework. Much of the current discourse is focused on land-based sustainability efforts and carbon emissions, with many initiatives being led by the public sector. Thailand has begun to recognize the importance of supporting the growth of private sector investments as indicated by recent actions implemented by government agencies. However, most of these initiatives are directed toward general sustainability, and not specifically toward the blue economy.

The Government of Thailand has indicated interest in supporting the development of blue initiatives within relevant sectors. For example, under the National Strategy (2018–2037), Thailand sets out its key eco-friendly development goals for the next 20 years that include achieving sustainable maritime-based economic growth through its various government agencies.[63] The National Strategy cascades down to the operational priorities of other relevant government agencies.

The Appendix table summarizes key public sector policy actions and initiatives related to the development of the blue economy and sustainable finance in Thailand.

---

[63] Published on 8 October 2018 pursuant to Section 65 of the Thai Constitution and the National Strategy Act B.E. 2560. Under these two legislations, a special committee with the approval of the Cabinet and the Senate, has the duty to set the strategic direction of the nation and set sustainable goals, to be used as a framework for other implementation plans. Government of Thailand, Office of the National Economic and Social Development Board, National Strategy Secretariat Office. 2018. National Strategy 2018–2037 (Summary). https://www.bic.moe.go.th/images/stories/pdf/National_Strategy_Summary.pdf.

Thailand: E Smart Bangkok Mass Rapid Transit Electric Ferries Project.
The Project involves development and operation of about 27 fully electric ferries for mass public transport along the Chao Phraya River in Bangkok, Thailand. The service will be operated by E Smart and will operate on a 30km stretch of the Chao Phraya river in Bangkok and Nonthaburi province.

IV. BARRIERS TO SCALING UP PRIVATE SECTOR FINANCING

Private sector financing for blue investments in Thailand is nascent as financial institutions are yet to prioritize such investments relative to those in other sectors. The following section summarizes the key barriers to financing blue economy sectors identified through comprehensive research and interviews with stakeholders, including financial institutions, private sector companies, government agencies, and sustainable finance experts.

Limited Awareness Regarding the Strategic Importance of Blue Economy Sectors for Sustainable Development

While awareness about the importance of sustainability and the green economy is improving, knowledge regarding the strategic importance of blue economy sectors is relatively low in Thailand. For example, while there are robust initiatives such as the Sustainable Finance Initiatives by the Working Group on Sustainable Finance and various fiscal incentives for the BCG economy sectors, most do not specifically target the blue economy. This limited awareness extends to lenders, borrowers, and the public. Increasing public awareness is important as public pressure may spur financial institutions and corporations to prioritize sustainable business practices that strengthen the blue economy. Similarly, public pressure is also likely to encourage policymakers to implement legislation and incentives with greater urgency. There is also confusion about how the green economy and the blue economy are different. Developing distinct taxonomies and disseminating knowledge is important in furthering the targeted development required to make comprehensive and sustainable progress.

This barrier can be identified across all blue economy sectors in Thailand to varying degrees. Pollution control-related sectors, sustainable fisheries, and sustainable aquaculture are relatively more progressive. Various government initiatives and policies, such as the BCG economy model and the ban on single-use plastic bags, are playing a vital role in increasing awareness as well as directing capital flows toward sectors that are addressed in such initiatives and policies.

Slow Pace of Integration of Sustainability Targets into Core Business Activities for Lenders and Borrowers

To produce tangible outcomes, simply improving awareness about the strategic importance of blue economy sectors will not be enough. The integration of sustainability principles into core business activities of financial institutions and the private sector in Thailand is required but remains limited. For example, most banks and corporates have established internal sustainability teams, which are relatively knowledgeable about best practices related to sustainable internal business processes. However, there is often an integration challenge with core operations (e.g., product and credit officers and relationship managers). The business case for incorporating sustainability should be sufficiently prioritized and better articulated by management teams. Another example is related to product offerings from financial institutions. Many commercial banks offer sustainability-oriented lending products, but most of these products are geared toward renewable energy (especially solar photovoltaic projects, including both utility-scale projects and solar rooftops for green buildings or homes) and e-mobility (both for retail, e.g., electric vehicle ownership loan, and for infrastructure projects, e.g., electric mass transit system).

This barrier can be identified across all blue economy sectors in Thailand. However, in response to customer and stakeholder expectations, the sustainable fisheries and sustainable aquaculture sectors have made significant progress in integrating sustainability practices into core business activities across the value chain.

Lack of Monitoring and Environmental, Social, and Governance Compliance Tools

Technology-enabled data collection and portfolio monitoring systems help manage risk as they can identify red flags early, which then allows management to take corrective actions in a timely manner. Data gathered from such systems enable financial institutions to develop robust mechanisms for pricing and verification, and enable the creation of innovative and targeted products and solutions. Finally, better monitoring tools help regulators assess how environmental, social, and governance (ESG) risks, such as climate-related physical risks and transition risks, may impact the balance sheets of companies.

Similarly, sector-specific standards and certifications developed and issued by independent third parties allow financiers to have confidence in borrowers that are certified in this manner. While Thailand does not have a comprehensive sustainable finance taxonomy or a specific framework related to blue economy sectors, strong demand was identified from financial institutions for top-down guidance, particularly from the Bank of Thailand. Some Thai financial institutions, such as TMB Thanachart Bank (TTB), have developed proprietary frameworks for sustainable lending toward green and blue economy sectors. These frameworks adopt internationally accepted standards, such as the Green Bond Principles by the International Capital Market Association (ICMA).

While the operations of most potential borrowers are not yet aligned with these frameworks that makes it challenging to access bank loans, the sustainable aquaculture sector has made some progress in this area. For example, the Thai Agricultural Standard on Good Aquaculture Practices (GAqP) for Marine Shrimp Farm ensures that farming systems consider environmental integrity and social responsibility.[64] Local fish farmers must meet GAqP across all stages of farm practices in marine shrimp culture, such as culturing, harvesting, and post-harvest handling. As of October 2020, there are more than 9,000 GAqP-certified shrimp farms in the country, which account for around 70% of the total number of registered farms.[65] These GAqP standards help farmers raise product quality and productivity and ensure sustainable practices and environmental conservation.

Lack of Investment Readiness

To assess a project's investment readiness and financing potential, commercial banks review the merits of the project, the robustness of the business plan (including realistic and achievable financial projections), and how effectively the project's developers articulate both components. Some projects in the blue economy are lagging in these aspects. As a result, commercial banks are often reluctant to extend loans to blue economy projects, which are perceived to be higher risk for various reasons. For example, coastal resilience projects, such as the widening of Jomtien Beach in Pattaya to tackle coastal erosion and attract tourists to the city, are not commercially viable without government support such as subsidies and grants, hence not yet appropriate for private sector financing.[66] Similarly, projects in the ecosystem management and restoration sector, while providing co-benefits such as flood protection, often require blended finance structures and revenue streams from blue carbon financing and other credit systems to attract capital. In the area of marine renewables, the potential for energy generation from offshore wind farms and tidal wave energy is

---

[64] Government of Thailand, Ministry of Agriculture and Cooperatives, National Bureau of Agricultural Commodity and Food Standards. 2014. *Good Aquaculture Practices for Marine Shrimp Farm.*
https://www.acfs.go.th/standard/download/eng/GAP-FOR-MARINE-SHRIMP-FARM_EN.pdf.

[65] C. Boocharoen and A. K. Anal. 2021. Attitudes, Perceptions, and On-farm Self-reported Practices of Shrimp Farmers towards Adoption of Good Aquaculture Practices (GAP) in Thailand. *Sustainability*. 13 (9). 5194. https://doi.org/10.3390/su13095194.

[66] C. Pupattanapong. 2020. Work to Start on B586m Enlargement of Jomtien Beach. *Bangkok Post*. 12 June.
https://www.bangkokpost.com/thailand/general/1933848/work-to-start-on-b586m-enlargement-of-jomtien-beach.

untapped due to a lack of technical readiness. Studies have estimated that the whole Gulf of Thailand has a technical power potential of 7 GW and that the regions Prachuap Khiri Khan and Narathiwat have the most potential.[67] However, except for EGAT's floating solar hybrid projects, capacity from marine renewable energy is not included in Thailand's Power Development Plan (PDP).[68] Moreover, EGAT's planned 2.7 GW-capacity floating solar hybrid projects are all located on dams across the country and not offshore. An example is the 45 MW Sirindhorn Dam pilot project in Ubon Ratchathani province.

Additionally, due to limitations in resources required to develop sophisticated business plans and marketing documents, small and medium-sized enterprises (SMEs) in particular struggle to attract appropriate financing. To address these challenges, specialized government agencies and financial institutions, such as the Office of SMEs Promotion (OSMEP) and SME Development Bank, offer capacity building and entrepreneurship programs.

Industrial-scale companies in the sustainable fisheries and sustainable aquaculture sectors (such as Thai Union) are more investment-ready with lower risk profiles given their maturity and scale, and have unsurprisingly attracted the most institutional investment capital. For example, Thai Union has issued three sustainability-linked loans and two sustainability-linked bonds totaling more than B26 billion ($740.7 million, denominated in Thai baht, Japanese yen, and US dollar), thereby setting a strong precedent for more issuers in similar sectors. It is also notable that blue economy focused impact funds such as Mirova's Althelia Sustainable Ocean Fund,[69] which focuses on the sustainable fisheries and sustainable aquaculture sectors, and Circulate Capital's Circulate Capital Ocean Fund,[70] which focuses on the solid waste management and circular economy sector, incorporate blended finance structures and guarantees to de-risk investments and attract commercial capital.

Limited Availability of At-scale Projects

In general, commercial banks prefer to fund larger projects with lower perceived risk profiles. Thailand's blue economy sector borrowers are predominantly SMEs who account for more than a third of total loans disbursed making this barrier highly relevant to the blue economy. Non-infrastructure projects, which include coastal and marine tourism, sustainable fisheries, and sustainable aquaculture, tend to be relatively small scale. Similarly, existing carbon projects in the ecosystem management and restoration sector face uncertainty about returns and cash flow as they are largely dependent on the development of voluntary carbon markets and the adoption of blue carbon by stakeholders; these projects are also small scale. Commercial banks, therefore, tend to overcrowd the limited number of mostly renewable energy infrastructure projects. Separately, larger corporates launching blue projects prefer to remain conservative and keep such projects smaller in scale unless concessional capital is available from donors and other sources.

SMEs-Related Barriers

SMEs dominate the blue economy in Thailand as well as the broader Asia and the Pacific region.[71] Thus, barriers in lending toward Thailand's blue economy sectors are, to a certain extent, synonymous with barriers in lending toward Thai SMEs. These include the lack of collateral and nonexistent or poor credit history of borrowers, which contribute

[67] J. Waewsak, M. Landry, and Y. Gagnon. 2015. Offshore Wind Power Potential of the Gulf of Thailand. *Renewable Energy*. 81 (C). pp. 609–626. https://www.sciencedirect.com/science/article/abs/pii/S0960148115002517; P. Sawasklin, S. Saeung, and J. Taweekun. 2021. Study on Offshore Wind Energy Potential in the Gulf of Thailand. *International Journal of Renewable Energy Research*. 11 (4). pp. 1947–1958. https://www.ijrer.org/ijrer/index.php/ijrer/article/view/12213/pdf.

[68] Government of Thailand, Ministry of Energy. 2018. *Thailand's Power Development Plan 2018-2037 (Rev. 1)*. https://policy.asiapacificenergy.org/sites/default/files/Thailand's%20Power%20Development%20Plan%20%28PDP%29%20%282018–2037%29%20%28TH%29.pdf.

[69] Green Finance Institute. Althelia Sustainable Ocean Fund. https://www.greenfinanceinstitute.co.uk/gfihive/case-studies/sustainable-ocean-fund/.

[70] Circulate Capital. Investments. https://www.circulatecapital.com/investments.

[71] ADB. 2022c. *Financing the Blue Economy: Investments in Sustainable Blue Small-Medium Enterprises and Projects in Asia and the Pacific*. Manila. https://www.adb.org/sites/default/files/publication/806136/financing-blue-economy.pdf.

to increased perceived risk by lenders and limited access to loans. There is no exhaustive data about loan–to–value ratio for SMEs loans in Thailand, but it can be as low as 40%–60% for one product to as high as 100%–150% for another product.[72] Even with Thailand's relatively high SMEs lending portion (30% of total loans outstanding compared to Southeast Asian peers Viet Nam 21%, Indonesia 20%, Malaysia 16%, the Philippines 6%), this barrier remains challenging. Unsecured lending products as well as supporting financial and nonfinancial incentives (e.g., guarantees, expansion of eligible collateral types, and capacity building programs) are available in the market, but more support is needed to counteract this barrier. SMEs-dominated sectors such as sustainable fisheries, sustainable aquaculture, and marine and coastal tourism are the most affected by these barriers. Meanwhile, industrial-scale projects in infrastructure-related sectors, such as ports and shipping and marine renewable energy, are largely unaffected.

Climate Risks

The impacts of climate change ensure that climate risk will be an increasingly significant barrier in the future. There are two types of climate risks: physical risks and transition risks.[73] Physical risks are related to the physical impact of climate change and may have financial implications for organizations, such as direct damage to assets and indirect impacts from supply chain disruption. Acute physical risks are event-driven (e.g., extreme weather events, such as cyclones, hurricanes, heat or cold waves, or floods) and chronic physical risks refer to longer-term shifts in climate patterns (e.g., sustained higher temperatures, sea level rise, changing precipitation patterns) that may cause a rise in sea levels or chronic heat waves. Transition risks are related to how an entity mitigates and adapts to transitioning to a lower-carbon economy. Transition risks pose higher financial and reputational costs from policy changes that push toward decarbonization (e.g., carbon tax, carbon disclosure), technological development and deployment that displace old technology (e.g., renewable energy), and shifting market preferences (e.g., more sustainability-conscious customers and investors).

Challenges posed by both types of climate risks toward projects in blue economy sectors include damage to real estate assets due to flooding (acute physical risk), biodiversity degradation due to coral bleaching (chronic physical risks), and reduced sales as customers demand more transparent supply chains for seafood products (transition risk). Blue economy sectors, such as coastal marine tourism, are affected by several of these challenges, making it difficult to attract funding from financial institutions.

This barrier is especially relevant for commercial banks that have adopted or are planning to adopt the Task Force on Climate-related Financial Disclosures (TCFD) framework which helps banks assess the impact of climate change on loan portfolios. Due to higher climate risks, blue economy projects may not meet the criteria for low climate risk investment opportunities. It is important to note, however, that blue economy projects incorporate economic, social, and environmental sustainability factors and, therefore, contribute to mitigation and resilience.

One of the shared action priorities in ESG Declaration by the Thai Banker's Association (TBA) is to develop monitoring and reporting systems in line with Thailand's regulatory frameworks and global disclosure standards.[74] The TCFD framework is explicitly recommended in the Bank of Thailand (BOT) policy paper issued in August 2022 on transitioning toward environmental sustainability in Thailand's financial landscape.[75] So far, Kasikornbank is the only Thai commercial bank to release a TCFD report.[76]

---

[72] *TMBThanachart Bank Public Company Limited*. A fact sheet on business credits with collateral. https://media.ttbbank.com/1/document/sbo/sales_sheet_so_smooth.pdf; *The Nation Thailand*. 2017. SCB Expands SME Market with 150% LTV Lending for up to 30 Years. 26 January. https://www.nationthailand.com/breaking-news/30305006.

[73] Task Force on Climate-related Financial Disclosure. 2017. *Recommendations of the Task Force on Climate-related Financial Disclosures*. https://assets.bbhub.io/company/sites/60/2020/10/FINAL-2017-TCFD-Report-11052018.pdf.

[74] BOT. 2022b. TBA Launches ESG Declaration, a Strong Collective Commitment to Expediting Sustainable Development toward Better and Greener Economy. 29 August. https://www.bot.or.th/en/news-and-media/news/news-20220829-2.html.

[75] BOT. Financial Landscape: Environmental Sustainability. https://app.bot.or.th/landscape/en/paper/sustainable/environment/.

[76] Kasikorn Bank. 2021. *KBank's TCFD Report 2021: Task Force on Climate-related Financial Disclosures*. https://www.kasikornbank.com/SiteCollectionDocuments/sustainable-development/pdf/kbank-tcfd-report2021-th.pdf.

Riverboats cruise past the Wat Arun, an important Buddhist temple and a well-known landmark along Bangkok's Chao Phraya River.

V. STRATEGIES FOR MOBILIZING CAPITAL

This investment report identifies and recommends six product design ideas and four policy actions and incentives that address financing barriers facing Thailand's blue economy sectors. Several of the product ideas address multiple barriers and can be implemented in the short-term while others require active collaboration between traditional financial institutions and catalytic capital providers. The recommendations on policy actions and incentives require dialogue with government agencies to be included in long term policy development.

Recommended Product Designs

Sustainability-linked Loans

Financial institutions in Thailand generally consider blue economy projects as high-risk due to a lack of investment and technological readiness, unproven business models, and scalability challenges. Therefore, the number of blue economy projects eligible for lending is limited. Given this context, commercial banks could review existing loan portfolios and identify eligible clients in relatively established, adjacent sectors instead of trying to limit loan disbursements solely toward new projects and clients. For example, solid waste and wastewater management are important business processes in sectors such as manufacturing and retail. Extending sustainability-linked loans (SLLs) to existing clients with lower repayment risk in these sectors could be a quick-win for financial institutions. Interest rates for SLLs are typically linked to the borrower's ability to meet sustainability targets, therefore companies are incentivized to achieve these targets to secure attractive pricing. Moreover, SLLs are fungible and do not need to be tied to specific projects, thus enabling banks to expand their loan portfolios and meet their sustainability financing targets.

More specifically, commercial banks can encourage the adoption of SLLs and incentivize early adopters by offering a pricing step-down mechanism whereby borrowers who meet project-specific sustainability targets can avail lower interest rates. Conversely, interest rate hikes through a step-up mechanism can be used to deter borrowers from failing to meet their sustainability targets. To support and accelerate issuance of SLLs by commercial banks, investors can provide guidance on setting and monitoring sustainability performance targets. In addition to providing catalytic capital at lower rates through the step-down mechanism, subscribers of the SLL could potentially earn upside from the surplus generated by the step-up mechanism.

Over the last few years, the sustainability-linked loan market has grown globally and in Thailand as well.[77] Notably, Thai Union issued SLLs and sustainability-linked bonds with performance targets that focus on transition, such as improving supply chain oversight improvement and reducing carbon intensity reduction.[78] The Bank of Ayudhya's (Krungsri) portfolio in 2021 includes SLLs to SCG Packaging Public Company Limited worth B5 billion ($142.5 million) with a 4-year tenor and to Thai Union Group PCL worth B6.5 billion ($185.2 million) with a 5-year tenor. In issuing SLLs, Krungsri follows best practices that are aligned with the core components of the Sustainability

---

[77] *Bloomberg NEF.* 2022. Sustainable Debt Issuance Breezed past $1.6 Trillion in 2021. 12 January. https://about.bnef.com/blog/sustainable-debt-issuance-breezed-past-1-6-trillion-in-2021/.

[78] *Thai Union.* 2021. Thai Union Launches Inaugural Sustainability-linked Loan. 16 February. https://www.thaiunion.com/en/newsroom/press-release/1292/thai-union-launches-inaugural-sustainability-linked-loan.

Linked Loan Principles, issued by the Asia Pacific Loan Market Association (APLMA), Loan Market Association (LMA), and Loan Syndications and Trading Association (LSTA). Similarly, a sustainability-linked bond issued by Indorama Ventures aims to reduce GHG emissions, improve recycling, and utilize more renewable electricity.[79]

Guarantees

A guarantee component can help de-risk blue economy projects by protecting banks and investors in cases of default. For example, the development risk of an offshore wind project is high as it requires project developers to estimate resource potential in order to create a realistic business plan. Investors seeking to catalyze these projects can act as the first-loss taker for the development cost in the case of insufficient resource potential or enable minimum take-or-pay payment for the off-take agreement. By assuming these roles, investors lower the risk profile of a project and, as a result, could increase the willingness of commercial banks to extend loans.

For SMEs, guarantees may be required to encourage commercial banks to extend loans without strict collateral and credit history requirements. In Thailand, the SMEs guarantee component is aligned with Thai Credit Guarantee Corporation's (TCG) offerings.[80] TCG is a state-owned specialized financial institution governed by the Ministry of Finance with the mandate to improve financial inclusion and business capabilities for SMEs by providing credit guarantees as well as technical assistance. In 2021, the amount of guarantee that TCG approved reached B246 billion ($7 billion) from more than 160,000 new cases with an average ticket size of around B1 million ($28,490).[81]

Technical Assistance

Technical assistance (TA) grants can also de-risk projects that lack investment readiness by providing support to facilitate the preparation, financing, and execution of blue economy projects. Such support strengthens capacities and promotes resource efficiency. Development organizations often combine TA and advisory services with a guarantee component to ensure the bankability of viable projects.[82] For example, the development risk of a mangrove restoration carbon project is high, as it requires project developers to confirm the eligible carbon credit potential even before preparing a business plan. Investors seeking to catalyze these projects can provide TA funding to pay for the project's initial development activities and provide project advisory services.

For SMEs, investors can play the role of business incubators and accelerators by providing technical capacity building programs to borrowers with the goal of improving business models as well as sustainable business practices. For example, in lending toward projects in the sustainable fisheries sector, government programs or development agency support could be structured to provide fishermen with technical training and know-how to improve productivity and catch yield, and to fish sustainably. The technical assistance financing component is also becoming increasingly common for forestry carbon projects and geothermal power plants. Another example is the Government of Indonesia's revolving fund for marine and fisheries SMEs (LPMUKP) which supports SMEs financing with capacity building programs.[83]

In Thailand, OSMEP's activities to promote and share ESG and BCG guidelines for SMEs help de-risk the development and implementation of sustainable projects. The Department of Agriculture (DOA) also provides local governments and farmers with TA for green projects and is likely to provide TA to help de-risk the execution of blue economy projects.

---

[79] *Indorama Ventures.* 2021. Indorama Ventures Issues Thb 10 Billion Sustainability-Linked Bond Driving Climate Action and Sustainable Production. 3 November. https://www.indoramaventures.com/en/updates/other-release/1840/indorama-ventures-issues-thb-10-billion-sustainability-linked-bond-driving-climate-action-and-sustainable-production.

[80] OECD. 2022. *Financing SMEs and Entrepreneurs 2022: An OECD Scoreboard: Thailand.* https://www.oecd-ilibrary.org/sites/b854dc2c-en/index.html?itemId=/content/component/b854dc2c-en.

[81] Thai Credit Guarantee Corporation. 2022. *Annual Report 2021.* https://www.tcg.or.th/uploads/file/221010103026gRHE.pdf.

[82] OECD. 2021. The Role of Guarantees in Blended Finance. *OECD Development Co-operation Working Papers No. 97.* https://www.oecd-ilibrary.org/docserver/730e1498-en.pdf?expires=1669175087&id=id&accname=guest&checksum=E1E0557F264D2536E1012FE279550A76.

[83] Climate Policy Initiative. 2022. Indonesia Blue Finance Landscape. https://www.climatepolicyinitiative.org/id/publication/indonesia-blue-finance-landscape/.

Blue Carbon Finance

Projects in blue economy sectors with strong public interest, such as ecosystem management and restoration and coastal resilience, do not have established commercial business models and returns are usually based on monetizing future cost savings potential. As a result, funders are limited to government agencies and nongovernment organizations, creating a bottleneck that depends on political will. In such projects, carbon credits can provide a potential revenue stream; enabling blue carbon finance could help improve the risk profile and catalyze capital from commercial investors. For example, mangrove restoration projects located in public coastal areas usually quantify returns based solely on future cost savings potential for the fiscal budget. While this is important from the government's point of view, it is not directly relevant for financial institutions. A revenue stream from carbon credit sales could make such projects more attractive for more traditional financial institutions, such as commercial banks.

Thailand has sizeable blue carbon potential and is well-positioned to identify carbon projects along its coastline. Investors can play a vital role in establishing structured incubators and accelerators for both public and private projects to improve their bankability. After a few successful pilot projects, commercial banks are likely to be more comfortable in extending loans toward similar projects (Box 1).

The discussion and consensus around blue carbon is relatively less established than green carbon, which is carbon sequestered by land ecosystems such as forests and peatlands. Verra, the global leader in voluntary GHG emissions reduction certification, has furthered the discussion with the release of its first blue carbon methodology in 2020 and the registration of its first blue carbon conservation project.[84]

Community-based Revolving Loans

SMEs loan sizes tend to be small commensurate with the borrowing capacity and balance sheet strength of the borrowers. To create scale in the SMEs loan portfolio in a cost-effective manner, commercial banks could partner and collaborate with existing cooperatives and aggregate their loan portfolios (Box 2). Banks can leverage the networks of cooperatives to disburse as well as monitor loan portfolios. In the blue economy context, it is also equally important to link the provision of credit to sustainable business practices. For example, instead of disbursing loans to individual fishermen, commercial banks could instead establish a revolving loan fund for a fishing village cooperative, which disburses the loans to fishermen, monitors the loans' repayment schedules, and provides regular updates to the lender. To align with the interest of commercial banks and ensure that the funds are responsibly spent, the revolving fund can only issue new loans to another project once the previous loan is repaid. This approach manages default risk by enabling a self-selecting process that only provides repeat loans to the most creditworthy borrowers.

Combining such a loan product with a TA component can also address market needs. Most local fishermen are not equipped with the technical capacity or know-how on sustainable fishing practices. By implementing a capacity building program, investors and commercial banks could manage the default rates at an acceptable level, while simultaneously achieving positive impact on the health of the ocean. In Thailand, market leaders have played a role in building the capacity of their suppliers. For example, CP Group has worked in 22 local fishing communities to promote environmental projects and R&D, and to use technology to help communities manage the marine environment and conserve resources.[85]

---

[84] *Verra*. 2020. First Blue Carbon Conservation Methodology Expected To Scale Up Finance For Coastal Restoration and Conservation Activities. 9 September. https://verra.org/first-blue-carbon-conservation-methodology-expected-to-scale-up-finance-for-coastal-restoration-conservation-activities/; Verra. 2021. Verra Has Registered Its First Blue Carbon Conservation Project. 10 May. https://verra.org/press-release-verra-has-registered-its-first-blue-carbon-conservation-project/.

[85] Charoen Pokphand Group Global. Ecosystem and Biodiversity Protection. https://www.cpgroupglobal.com/en/sustainability/home-living-together/ecosystem-and-biodiversity-protection.

Box 1: Carbon Credits in Thailand

A carbon credit is a tradable permit or certificate that allows the holder of the credit to emit a fixed amount of 1 ton per credit of carbon dioxide or other greenhouse gas (GHG). Carbon credit markets or trading originated from the notion that a market mechanism would create a monetary incentive for companies to reduce GHGs and the effects of global warming.[a]

Broadly, carbon markets can be categorized into two types: compliance or voluntary. The compliance market is a mandatory and regulated marketplace in which a fixed number of carbon credits are issued per company and per year. Since it is mandatory and regulated, companies must fulfill them. Companies that emit less than their limit may resell carbon credits in the corresponding carbon market. In comparison, the voluntary, self-governed market is neither mandatory nor enforced. Organizations or individuals with operations that generate carbon offsets can issue and sell credits to companies or individuals who want to measurably decrease the amount of carbon they emit.[b]

In 2021, the Bangchak Group, Kasikornbank, BTS Group, and CP Group in Thailand formed a carbon exchange group, namely 'the Carbon Markets Club' to build a framework based on the European Union and the People's Republic of China markets, where greenhouse gas emitters can buy carbon credits to offset emissions.[c]

In September 2022, Thailand launched its first carbon credit platform, marking a major step toward the country's goal to achieve carbon neutrality and combat climate change. The new carbon credit platform, called FTIX, will be operated by The Federation of Thai Industries, which comprises about 12,000 private companies across 45 sectors.[d] It will allow firms and government agencies to buy and sell carbon credits and track their emissions on an online dashboard, incorporating the government's existing voluntary emission program. Prior to the launch of the platform, only over-the-counter trading existed, in which developers of Clean Development Mechanism projects and countries listed in Annex I to the United Nations Framework Convention on Climate Change could trade credits through delegates, financial funds, and brokers.[e]

[a] United Nations Development Programme. 2022. What Are Carbon Markets and Why Are They Important? 18 May. https://climatepromise.undp.org/news-and-stories/what-are-carbon-markets-and-why-are-they-important#:~:text=In%20a%20.
[b] Carbon Offset Guide. Mandatory & Voluntary Offset Markets. https://www.offsetguide.org/understanding-carbon-offsets/carbon-offset-programs/mandatory-voluntary-offset-markets/.
[c] BTS Group. 2021. BTS Group Jointly Establishes the Carbon Markets Club, Promoting Carbon Credit Trading to Help Reduce Greenhouse Gases Towards a Net Zero Society. 29 June. https://www.btsgroup.co.th/en/update/news-event/688/bts-group-jointly-establishes-the-carbon-markets-club-promoting-carbon-credit-trading-to-help-reduce-greenhouse-gases-towards-a-net-zero-society.
[d] The Nation Thailand. 2023. Carbon Credit Trading Opens on FTIX Platform Tomorrow. 15 January. https://www.nationthailand.com/thailand/economy/40024038.
[e] K. Bunjongsiri. 2019. The Overview of Carbon Credit Market in Thailand. *SAU Journal of Science & Technology*. 5 (2). pp. 1–9. https://ph01.tci-thaijo.org/index.php/saujournalst/article/view/184556.

Source: ADB based on AWR Lloyd.

Revolving loan funds are used worldwide for businesses that are too small for a loan from commercial banks. A notable example is the California Fisheries Fund (CFF), which has been providing steady, accessible capital to California's harbor communities.[86] Historically, banks and credit unions have been reluctant to extend loans to the fishermen in these communities due to marginal profit expectations and the general unpredictability of the fishing industry. Furthermore, the CFF model is unique because it allows fishing boats, permits, and quotas to be pledged as collateral.

Box 2: Fishing Cooperatives in Thailand

Fishing cooperatives are among one of the seven types of cooperatives in Thailand—including agricultural cooperatives, credit union cooperatives, and land settlement cooperatives—monitored and promoted by the Cooperative Promotion Department, under the Ministry of Agriculture and Cooperatives.

Fishing cooperatives are set up to aggregate individual members involved in fishing activities both in freshwater and seawater and achieve the following objectives:[a]

- sell aquatic animals and processed products of its members in the market,
- sell fishing equipment and other necessities to members at reasonable prices,
- provide access to low-interest loans to members,
- accept deposits from members,
- enhance knowledge and technical and business capacity through trainings, and
- strengthen livelihoods of members by providing welfare during crises.

Types of fishing cooperatives include:
  i **Sea fisheries cooperatives.** Members include large fishermen who fish outside Thailand's territorial waters, medium fishermen within its territorial waters, and small fishermen within 3,000 meters offshore
  ii **Brackish water fisheries cooperatives.** Members are involved in the cultivation of shrimp, aquatic animals, and brackish water farming
  iii **Freshwater fisheries cooperatives.** Members are involved in the farming of freshwater fish and other aquatic animals

The Pissanu Fishery Cooperative—Thailand's first fisheries cooperative—was established in 1949 in the Phitsanulok province, with 54 members.[b] The cooperative introduced new fishery techniques and promoted the preservation of aquatic animals, alongside other activities in efforts to support land allocation, marketing, the processing of aquatic animals, and receiving concessions.

IUU = illlegal, unreported, and unregulated.
[a] Government of Thailand, Ministry of Agriculture and Cooperatives, Cooperative Promotion Department. 2008. *Fishing Cooperatives*. https://www.cpd.go.th/cpden/images/FisheriesCooperatives.pdf.
[b] Cooperative League of Thailand. Fishing Cooperatives. https://www.cltcoop.com/17624383/fishery-cooperatives.
Source: ADB based on AWR Lloyd.

[86] California Ocean Protection Council. 2018. *California Fisheries Fund.* 7 September. https://www.opc.ca.gov/2010/01/california-fisheries-fund/.

Insurance

Projects in the blue economy are vulnerable to the physical risks of climate change. Given the intensifying nature of these risks and the need to adapt and mitigate impacts on commercial operations, insurers could be encouraged or incentivized to develop insurance products that help mitigate such risk for lenders and investors. Insurers can potentially collaborate with financiers to develop such products and create tools to calculate risk exposure.

A recent example is coral reef insurance (Box 3), which provides funds to protect, restore, and repair damaged reefs in the immediate aftermath of hurricanes or tropical storms. Other notable blue insurance products include ADB's $3.8 million support for the development of financial risk management products, such as coral reef insurance, in Fiji, Indonesia, the Philippines, and Solomon Islands.[87]

In Thailand, a relevant example of climate risk insurance is the rice insurance plan for farmers.[88] This calamity-based insurance scheme was established for farms located in vulnerable areas and covers damage caused by disasters triggered by natural hazards including heavy rains, flooding, drought, storms, cold, hail, fires, and wild elephants. The payout covers total losses as determined through government inspections. The Thai government and Bank of Agriculture and Agricultural Cooperatives subsidizes the premium, with the latter playing the role of distributing the premiums. Future improvements in the scheme include using remote sensing technology to support claims declaration and assessment and creating a parametric insurance scheme (e.g., using weather-based index).

Box 3: Coral Reef Insurance

In November 2022, global advisory firm Willis Towers Watson (WTW) and The Nature Conservancy (TNC) joined forces to announce the first coral reef insurance policy in the United States. This policy will provide coverage for rapid coral reef repair and restoration in Hawaii following damage from hurricanes or tropical storms. This parametric insurance is triggered at windspeeds of 50 knots (92.6 kilometers per hour) if sufficiently close to reefs and can provide pay-outs up to $2 million to allow rapid reef repair and restoration after storm damage. The policy will be in place through the 2023 hurricane season.

In 2019, TNC launched the world's first reef insurance policy to protect against hurricane damage in Quintana Roo, Mexico. After Hurricane Delta in October 2020, a cash pay-out from the policy was used to help repair the damage caused by the storm. Since then, the Mesoamerican Reef Fund, working in collaboration with WTW's Climate and Resilience Hub, to design a new parametric insurance program across the Mesoamerican Reef. The Hawaii policy adds tropical storms as a covered event, because tropical storms can cause significant damage without making landfall.[a]

Tropical storms and hurricanes, which are increasing in intensity due to climate change, are a major short-term threat to coral reefs. Research shows that severe hurricanes can cause a 50% or more loss of live coral cover, and the loss of just one meter of reef height could result in a doubling of the cost of damage. Healthy, intact reefs can reduce up to 97% of wave energy and are the islands' first line of defense during storms.[b]

[a] A. Rolt. 2022. The Nature Conservancy Debuts New Hawaii Coral Reef Insurance Plan. Green Biz. https://www.greenbiz.com/article/nature-conservancy-debuts-new-hawaii-coral-reef-insurance-plan.

[b] TNC. 2022. The Nature Conservancy Announces First-ever Coral Reef Insurance Policy in the US. 21 November. https://www.nature.org/en-us/newsroom/first-ever-us-coral-reef-insurance-policy/.

Source: ADB based on AWR Lloyd.

---

[87] ADB Approves $3.8 Million Support for Development of Coral Reef Insurance News release. https://www.adb.org/news/adb-approves-3-8-million-support-development-coral-reef-insurance#:~:text=MANILA%2C%20 PHILIPPINES%20(30%20September%202022,Southeast%20Asia%20and%20the%20Pacific.

[88] C. Theparat. 2022. Panel Okays B1.92bn Rice Insurance Plan. *Bangkok Post*. 28 April. https://www.bangkokpost.com/business/2301478/panel-okays-b1-92bn-rice-insurance-plan.

Recommended Policy Actions and Incentives

The recommended policy actions and incentives detailed below aim to create the enabling environment required to catalyze private sector participation and increase capital flows toward the blue economy in Thailand. These recommendations complement and address barriers not addressed by product design suggestions. To ensure adoptability, effectiveness, and efficiency, they are aligned as much as possible with existing strategies, policies, and initiatives of the Government of Thailand and related entities.

Develop a Sustainable Finance National Taxonomy

A sustainable finance taxonomy refers to a classification system of environmentally sustainable economic activities. A national unifying taxonomy that incorporates blue economy sectors will improve market clarity and raise awareness regarding the strategic importance of the blue economy in Thailand. In turn, the heightened awareness can: (i) enable policymakers to implement supporting regulatory frameworks, (ii) enable companies to develop climate-friendly projects in line with investor expectations, (iii) bolster investor confidence and channel capital toward sustainable blue investments, and (iv) help regulators to monitor and measure capital flows toward the blue economy and to refine policies further.

In August 2021, the Bank of Thailand, alongside other key players, published its intention to develop a sustainable finance taxonomy tailored to the Thai market. The Sustainable Finance Initiatives report by the Working Group on Sustainable Finance BOT highlights the importance of incorporating locally relevant and impactful SMEs-related activities in the taxonomy.[89] While the first version of the taxonomy will focus on climate change mitigation, the BOT intends to include the sustainable use of ocean and marine resources.[90] Similarly, the Securities and Exchange Commission (SEC) is working to develop policies to regulate the issuance of sustainability-linked bonds.[91] This will complement existing regulations on ESG bonds and allow for more inclusive and diversified access to green and blue bonds to nonconventional issuers. It will also be relevant for SMEs intending to transition toward sustainability.

In 2021, TTB, one of Thailand's most active commercial banks in the blue economy, developed its own Green and Blue Bond Framework.[92] This framework was developed in accordance with the internationally recognized Green Bond Principles issued by the ICMA and is a reflection of the support a national taxonomy will receive from the private sector.

Create and Nurture a Better Data Environment

A robust, reliable, and accessible data environment is required to develop efficient, investment-ready blue projects aligned with the national taxonomy, government policies, and investor expectations. An improved data environment will enable companies to more accurately determine sustainability metrics and meet disclosure requirements of globally accepted frameworks, such as the TCFD and the Global Reporting Initiative (GRI). Examples of these metrics and data points include GHG emissions, use of resources, diversity, training and capacity development, and labor rights. Blue economy specific examples include offshore wind potential, ballast water management, volume of wastewater treated, blue carbon sequestration estimates, and the percentage of certified sustainable fisheries. In Thailand, the BOT uses the TCFD as the benchmark among international frameworks that consider climate change (footnote 76). This is now complemented by the Taskforce on Nature Related Financial Disclosures.

---

[89] Working Group on Sustainable Finance. 2021. *Sustainable Finance Initiatives for Thailand.* https://www.sec.or.th/TH/Documents/KnowledgeBase/SustainableFinanceInitiativesforThailand.pdf.

[90] BOT. 2022a. Directional Paper on Transitioning Towards Environmental Sustainability under the New Thai Financial Landscape. BOT Press Release No. 43/2022. 23 August. https://www.bot.or.th/en/news-and-media/news/news-20220823.html.

[91] Government of Thailand, SEC. 2021. *Strategic Plan 2020–2023.* p. 29. https://www.sec.or.th/TH/Documents/strategicplan/strategicplan-2564-2566.pdf.

[92] TTB. 2022. *TTB Green and Blue Bond Framework.* https://media.ttbbank.com/1/ir/green-blue-bond/ttb-green-blue-bond-framework-en.pdf.

Separately, to strengthen the confidence of financial institutions toward sustainable projects, the SEC is promoting the use of third-party verification to evaluate and validate ESG disclosures published in the 56-1 One Report, the combined disclosure of annual registration statements and annual reports regulated by the SEC. The 56-1 One Report aims to ensure that ESG considerations are factored into long-term business strategies and embedded in operations. While Thailand has not issued a list of eligible third-party reviewers, the BOT has announced its intention to progress ESG-related initiatives by establishing a central agency in coordination with the Thailand Greenhouse Management Organization and the SEC.

Internationally, the United Nations Environment Programme Finance Initiative Sustainable Blue Economy Finance Initiative works with the finance community to support the implementation of the Sustainable Blue Economy Finance Principles. This initiative provides frameworks (which can guide data collection and disclosure best practices) to ensure investment, underwriting, and lending activities are aligned to the UN Sustainable Development Goal 14. In addition, the United Nations Global Compact Ocean Stewardship Coalition supports companies to develop blue investments.

Promote Private Participation through Financial Incentives

Redirect unsustainable subsidies. SDG 14.6 charges the World Trade Organization with convening countries to change fisheries subsidies that incentivize overfishing and eliminate those that contribute to IUU fishing. Thailand has a significant opportunity to redirect subsidies in the fishing sector to incentivize activity in support of the blue economy.

Feed-in tariffs. Thailand has a pilot project testing the feed-in tariff (FiT) scheme to promote the nation's solar energy generation capacity. Pursued under the 20-year renewable energy development plan, the Minister of Energy mandated the purchase of approximately 200 MWh of energy from rooftop solar projects throughout the nation at the FiT rates averaging around B6.6 ($0.19)/unit for 25 years counting from the commercial operation date. Thailand could extend the FiT scheme for floating solar projects, or other renewable energy sources related to the blue economy.

Tax holidays. To further encourage private investments in solar projects, BOI Thailand specified the following measures: (i) corporate income tax exemption, whereby companies manufacturing solar cells or its parts may be eligible for an 8-year tax exemption; and (ii) import duty waiver, whereby companies importing new machinery, raw materials, and manufacturing tools are exempt from paying import duties. As of 2017, approximately B2 billion ($57 million) worth of capital has benefited from these special privileges and benefits from the BOI Thailand.[93] Similar tax holidays can be extended to the blue economy, such as for the development of offshore wind farms.

Integrate Public Sector Efforts

The blue economy cuts across multiple sectors and requires integrated policies and networks to develop effective investments. Such policies can meaningfully contribute to a broad range of sustainable development outcomes and attract desired capital flows. For example, land-based pollutants such as garbage and sewage, agricultural run-off, and mining waste seriously affect marine ecosystems but are often not considered in terms of managing the sustainablility of marine-related productive activities.

Government initiatives in Thailand remain largely sectoral and often do not leverage the required cross-sectional expertise and coordination. Several financial institutions and corporates interviewed indicated that clearer direction and ownership of existing government policies, along with regulatory support for ministries from the central government, are required to drive capital flows into the blue economy. For example, the MNRE reports that initiatives to issue bonds to raise capital for the restoration of forest areas under 2012–2016 Implementation Plan could not proceed due to a lack of coordinated supporting legislation. Clear ownership and collaboration within the government is required for the robust development of Thailand's blue economy.

[93] Government of Thailand, Ministry of Energy, Department of Alternative Energy Development and Efficiency. 2018. *Final Report on the Status of Electricity Generation from Solar in Thailand.* PDF_PVstatus2561.pdf (dede.go.th).

VI. ROAD MAP FOR SCALING UP FINANCING TOWARD THE BLUE ECONOMY

The road map developed for scaling up private sector financing toward the blue economy in Thailand categorizes sectors into four different groups with varying prioritization levels and action plans. While some groups require short-term actions, others require long-term policy developments.

The road map was prepared by assessing the following factors relevant to blue economy sectors:

A. **Sector attractiveness.** This is determined by ranking blue economy sectors based on five key factors.
B. **Financial institution interest.** This is determined by identified interest from financial institutions.
C. **Prioritization of sectors.** This involves plotting sectors into a four-quadrant matrix with related actions.

Attractiveness of Blue Economy Sectors to Financial Institutions

The attractiveness of blue economy sectors is based on the following five factors: (i) current market size, (ii) growth potential, (iii) environmental impact, (iv) social impact, and (v) regulatory incentives.

As shown in Figure 4, the most attractive blue economy sectors for financial institutions are solid waste management, wastewater management, sustainable fisheries, coastal and marine tourism, and sustainable aquaculture.

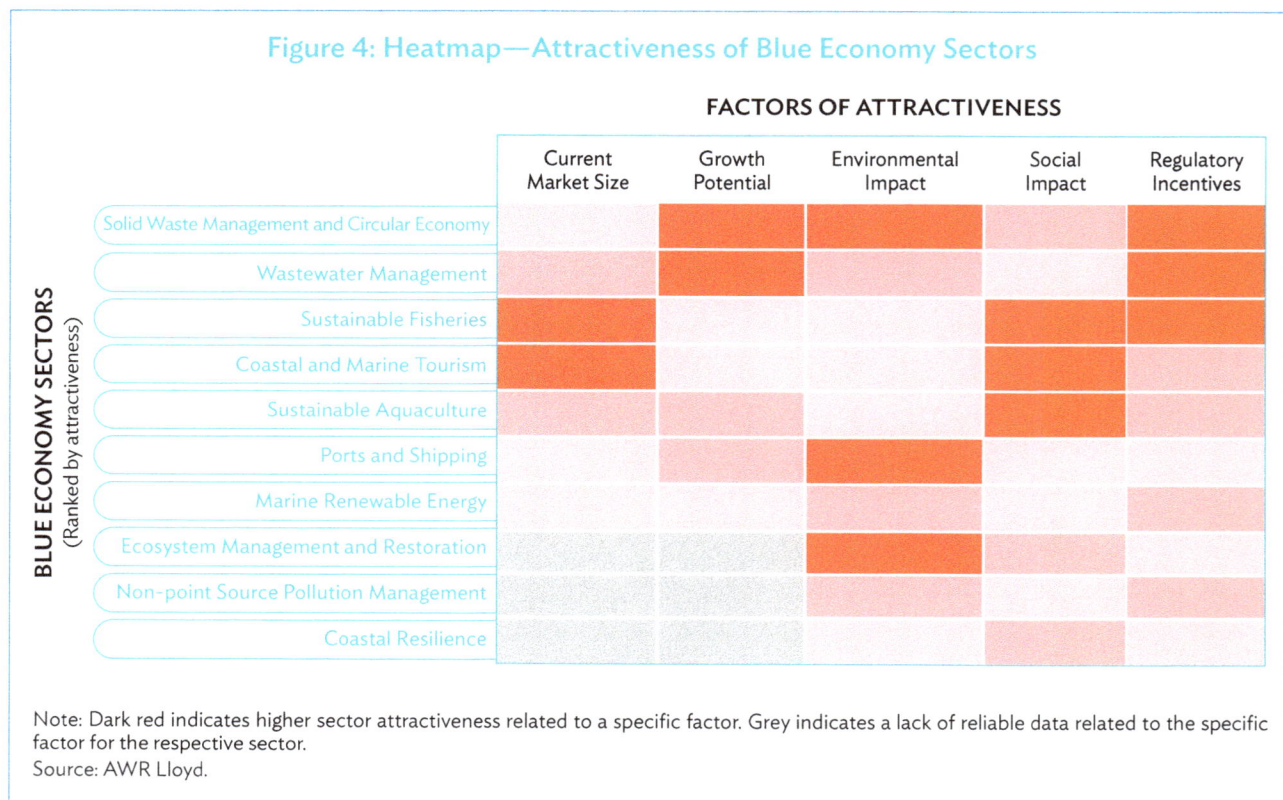

Figure 4: Heatmap—Attractiveness of Blue Economy Sectors

FACTORS OF ATTRACTIVENESS

BLUE ECONOMY SECTORS (Ranked by attractiveness)	Current Market Size	Growth Potential	Environmental Impact	Social Impact	Regulatory Incentives
Solid Waste Management and Circular Economy					
Wastewater Management					
Sustainable Fisheries					
Coastal and Marine Tourism					
Sustainable Aquaculture					
Ports and Shipping					
Marine Renewable Energy					
Ecosystem Management and Restoration					
Non-point Source Pollution Management					
Coastal Resilience					

Note: Dark red indicates higher sector attractiveness related to a specific factor. Grey indicates a lack of reliable data related to the specific factor for the respective sector.
Source: AWR Lloyd.

Interest from Financial Institutions

The heatmap in Figure 5 assesses the interest of Thailand's banks (categorized by size) and SFIs in blue economy sectors. The heatmap reveals that blue economy sectors that rank highly in attractiveness also receive the strongest interest from Thai financial institutions. It is notable that the coastal and marine tourism sector has not been a focus for Thai financial institutions despite ranking quite highly in terms of sector attractiveness. Also, lack of interest in non-point source pollution management is likely due to a nomenclature issue, whereby it is included in other pollution related sectors.

Figure 5: Heatmap—Interest from Financial Institutions

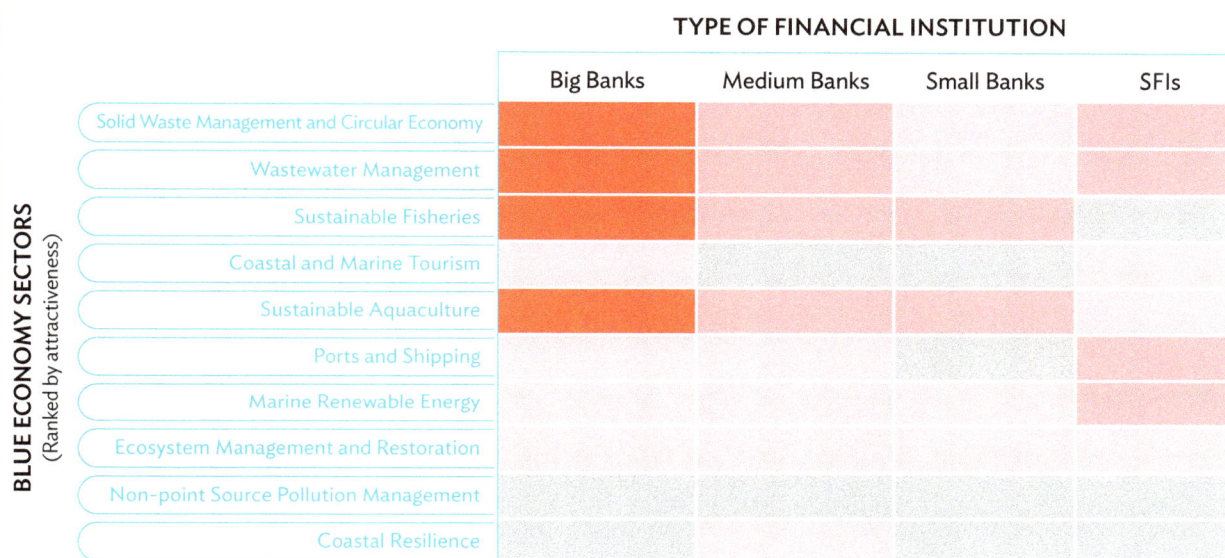

TYPE OF FINANCIAL INSTITUTION

BLUE ECONOMY SECTORS (Ranked by attractiveness)	Big Banks	Medium Banks	Small Banks	SFIs
Solid Waste Management and Circular Economy				
Wastewater Management				
Sustainable Fisheries				
Coastal and Marine Tourism				
Sustainable Aquaculture				
Ports and Shipping				
Marine Renewable Energy				
Ecosystem Management and Restoration				
Non-point Source Pollution Management				
Coastal Resilience				

SFI = special financial institution.
Notes:
1. Dark red indicates higher support from financial institutions to each subsector in the blue economy in Thailand. Grey indicates a lack of reliable data to accurately assess interest levels.
2. There are 29 commercial banks in Thailand and seven state-owned SFIs. Asset sizes for commercial banks is as of July 2022 with the following categorization: (i) big banks: assets more than B1 trillion ($28.5 billion), (ii) medium banks: assets between B200 billion ($5.7 billion) and B1 trillion ($28.5 billion), and (iii) small banks: assets less than B200 billion ($5.7 billion). Asset sizes for SFIs is sourced from latest annual reports.

Source: AWR Lloyd.

Prioritization of Blue Economy Sectors

As the final step in developing the private sector financing road map for the blue economy, each sector is plotted into a four-quadrant matrix as illustrated in Figure 6. Each quadrant represents the following groups: quick wins, hidden gems, stepping stones, and question marks. The X-axis represents interest levels from financial institutions and the Y-axis represents sector attractiveness. The results highlight the following:

1. **Quick Wins (top-right).** Sectors within pollution control, sustainable fisheries, and sustainable aquaculture rank the highest in sector attractiveness and interest from financial institutions.

2. **Hidden Gems (top-left).** The coastal and marine tourism sector ranks highly in terms of sector attractiveness but has not received strong interest from financial institutions.

3. **Stepping Stones (bottom-right).** Ports and shipping and marine renewable energy rank lower in terms of sector attractiveness but have received some interest from financial institutions.

4. **Question Marks (bottom-left).** Ecosystem management and restoration and coastal resilience rank the lowest in terms of sector attractiveness and interest from Thai financial institutions.

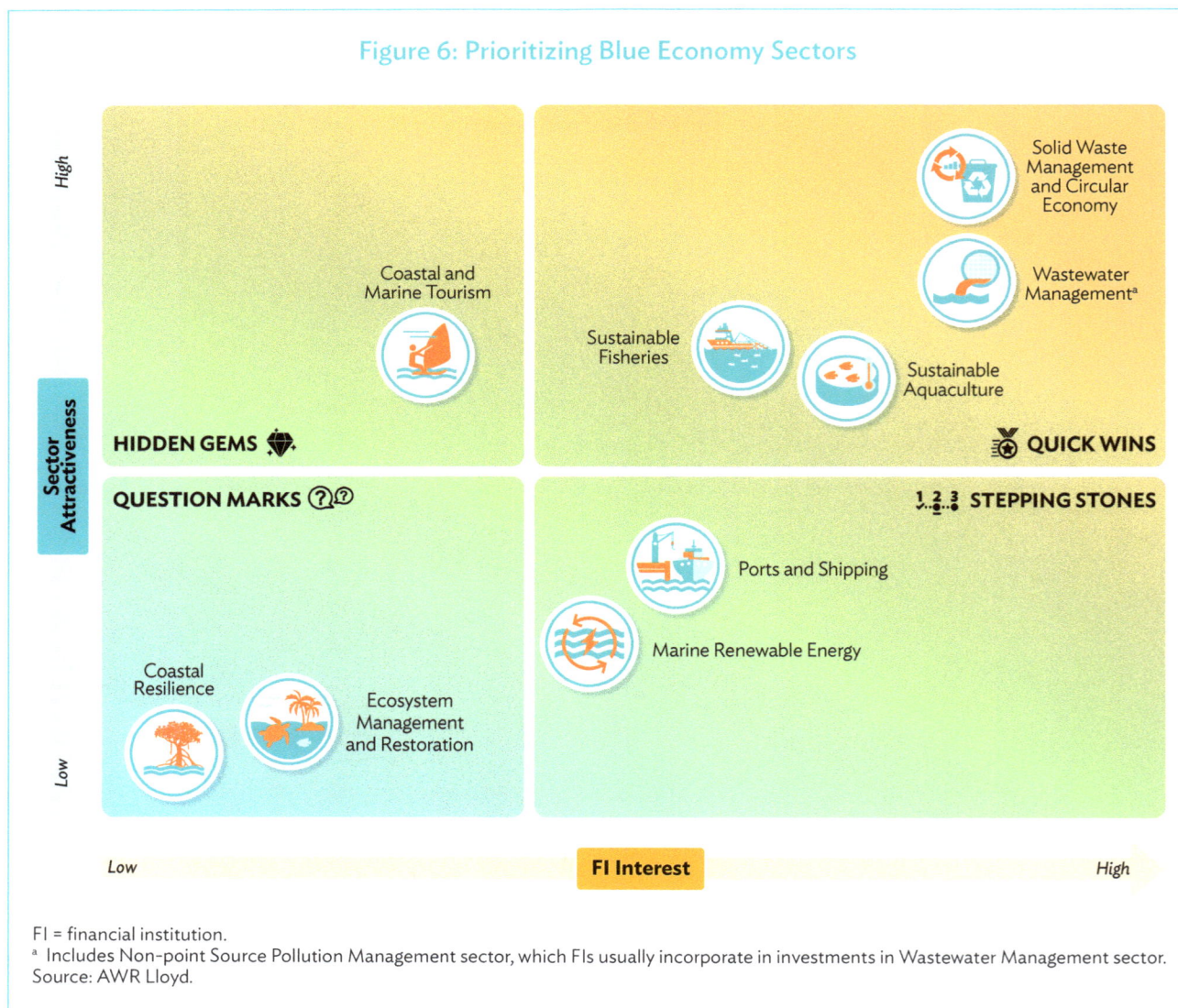

Figure 6: Prioritizing Blue Economy Sectors

FI = financial institution.
ª Includes Non-point Source Pollution Management sector, which FIs usually incorporate in investments in Wastewater Management sector.
Source: AWR Lloyd.

Road Map for Scaling Up Private Sector Financing for the Blue Economy in Thailand

The road map in Figure 7 summarizes barriers to financing and provides corresponding solutions for each blue economy sector.

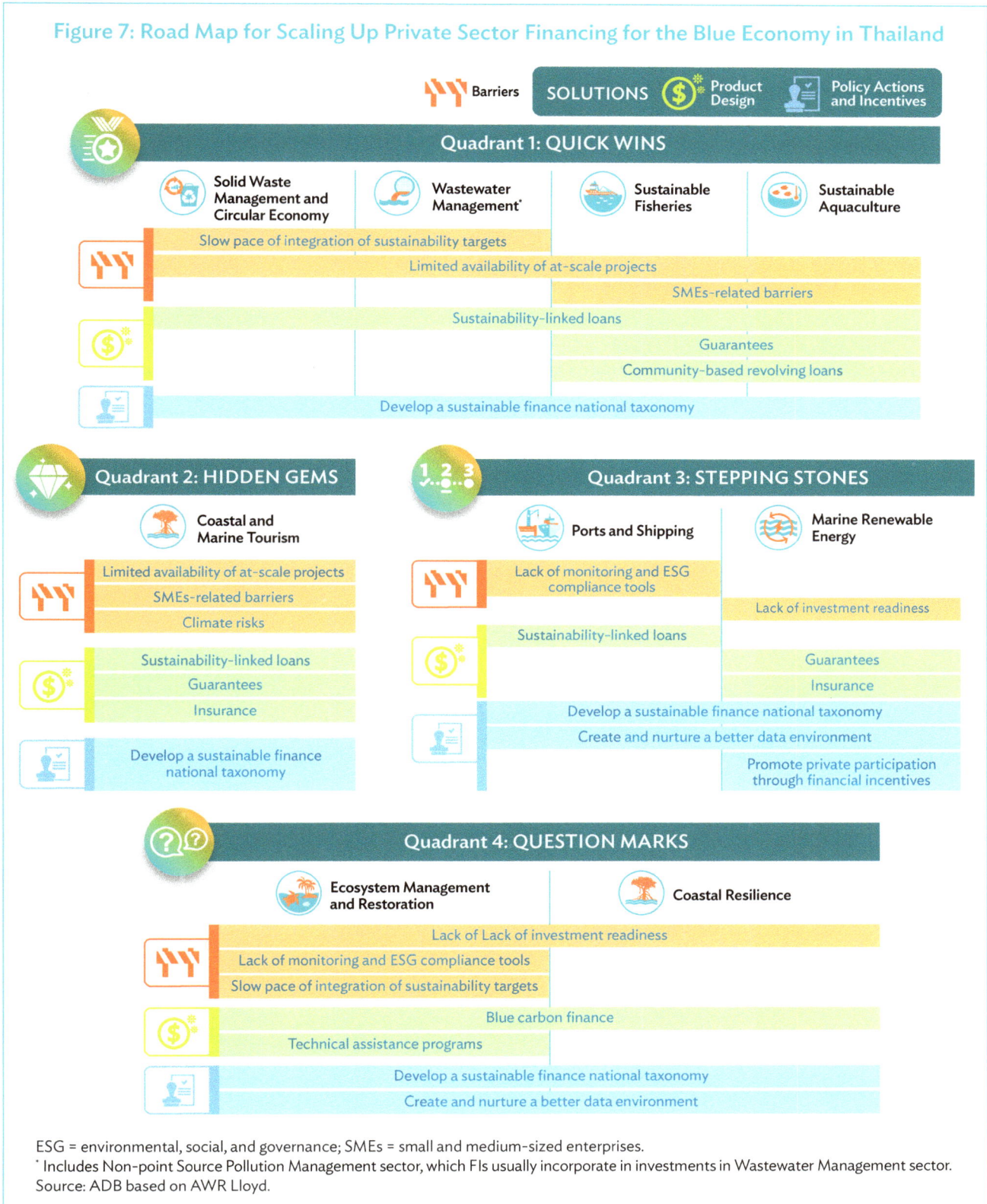

Figure 7: Road Map for Scaling Up Private Sector Financing for the Blue Economy in Thailand

Barriers | SOLUTIONS | Product Design | Policy Actions and Incentives

Quadrant 1: QUICK WINS

	Solid Waste Management and Circular Economy	Wastewater Management*	Sustainable Fisheries	Sustainable Aquaculture
Barriers	Slow pace of integration of sustainability targets			
	Limited availability of at-scale projects			
			SMEs-related barriers	
Product Design	Sustainability-linked loans			
			Guarantees	
			Community-based revolving loans	
Policy Actions	Develop a sustainable finance national taxonomy			

Quadrant 2: HIDDEN GEMS

	Coastal and Marine Tourism
Barriers	Limited availability of at-scale projects
	SMEs-related barriers
	Climate risks
Product Design	Sustainability-linked loans
	Guarantees
	Insurance
Policy Actions	Develop a sustainable finance national taxonomy

Quadrant 3: STEPPING STONES

	Ports and Shipping	Marine Renewable Energy
Barriers	Lack of monitoring and ESG compliance tools	
		Lack of investment readiness
Product Design	Sustainability-linked loans	
		Guarantees
		Insurance
Policy Actions	Develop a sustainable finance national taxonomy	
	Create and nurture a better data environment	
		Promote private participation through financial incentives

Quadrant 4: QUESTION MARKS

	Ecosystem Management and Restoration	Coastal Resilience
Barriers	Lack of Lack of investment readiness	
	Lack of monitoring and ESG compliance tools	
	Slow pace of integration of sustainability targets	
Product Design	Blue carbon finance	
	Technical assistance programs	
Policy Actions	Develop a sustainable finance national taxonomy	
	Create and nurture a better data environment	

ESG = environmental, social, and governance; SMEs = small and medium-sized enterprises.
* Includes Non-point Source Pollution Management sector, which FIs usually incorporate in investments in Wastewater Management sector.
Source: ADB based on AWR Lloyd.

The key takeaways from the road map include the following:

(i) **Sectors in pollution control** such as solid waste management and circular economy, wastewater management, and non-point source pollution management rank **high in attractiveness and interest from financial institutions in Thailand**. While several institutions have invested in this sector, these investments remain small and could be scaled up. Action plans should focus on expanding financing through recommended product designs.

(ii) **Sustainable fisheries and sustainable aquaculture sectors are attractive segments and have received considerable interest from financial institutions in Thailand**. Unlike in the pollution sector, activities in sustainable fisheries and aquaculture have been driven by a few large issuers with focused interest in these areas. There is potential for growth in this space by replicating viable models for smaller issuers and SMEs that dominate these sectors.

(iii) **Coastal and marine tourism is an attractive segment but has not been a focus area historically for Thai financial institutions**. Developing recommended product designs and pushing for regulations and incentives that alleviate barriers should be prioritized.

(iv) **Ports and shipping and marine renewable energy sectors have received limited interest from financial institutions** who should be encouraged to develop targeted financing strategies and products for these sectors.

(v) **Ecosystem restoration and management and coastal resilience sectors are generally noncommercial and consequently unattractive for financial institutions**. While these sectors provide critical co-benefits such as flood protection, they often require blended finance structures and revenue streams from blue carbon financing and other credit systems to unlock commercial value and attract capital.

Support from ADB and the ASEAN Catalytic Green Finance Facility

ADB has been a leader in promoting the blue economy through its $5 billion Action Plan for Healthy Oceans and Sustainable Blue Economies, guided by its Ocean Finance Framework. In 2020, it approved its first independently verified nonsovereign blue loan, the Indorama Ventures Regional Blue Loan Project, following the Blue Natural Capital Financing Facility's Blue Bond Guidelines, with an assurance report from DNV GL, and aligned with the Action Plan (Box 4). In 2022, ADB issued its first blue bond under its expanded Green and Blue Bond Framework and aligned with the ICMA. In June 2023, ADB signed a $44.2 million nonsovereign blue loan with PT ALBA Tridi Plastics Recycling Indonesia, an ALBA Group Asia company, to support the development, construction, and operation of a PET recycling facility in Central Java, Indonesia (Box 4).

Blue economy projects in Thailand and other member countries of the Association of Southeast Asian Nations (ASEAN) may be eligible for TA grants from the ASEAN Catalytic Green Finance Facility (ACGF) and for support from the Blue SEA Finance Hub. The ACGF is a $1.9 billion permanent facility under the ASEAN Infrastructure Fund (AIF), dedicated to accelerating green infrastructure investments in Southeast Asia. It supports ASEAN member governments to prepare and finance infrastructure projects that promote sustainability and contribute to climate goals. The ACGF is owned by ASEAN member governments and ADB and is administered by ADB.

The ACGF focuses on catalytic green and blue finance and leverages its resources to de-risk infrastructure projects to make them more bankable and attract various sources of funding, including private, institutional, and commercial funds. The ACGF offers access to (i) funds from the AIF, through two lending products with differentiated pricing based on the differing socioeconomic conditions of ASEAN countries; (ii) financing from nine ACGF partners; and (iii) TA from ADB and ACGF partners.

Box 4: Examples of ADB-Supported Blue Projects

Indorama Ventures Regional Blue Projects. In November 2020, the Asian Development Bank (ADB) provided nonsovereign financing to Indorama Ventures group, a global integrated polyethylene terephthalate (PET) manufacturer, to expand the company's PET recycling capacity in India, Indonesia, the Philippines, and Thailand under the Indorama Ventures Regional Blue Loan Project.[a] PET plastics are widely used in beverage bottles, hence ADB's financing will help promote a circular economy for recycled PET aligned with industry best practices.

The total loan package of $300 million comprises of $50 million from ADB, $50 million from the ADB-administered Leading Asia's Private Infrastructure Fund, $150 million from the International Finance Corporation, and $50 million from the Deutsche Investitions- und Entwicklungsgesellschaft mbH. As ADB's first internationally verified nonsovereign blue loan, the financing will be contingent on the borrower complying with ADB's environmental and social safeguard standards.

The project aligns with ADB's Action Plan for Healthy Oceans and Sustainable Blue Economies, which calls for the bank to grow its investments and technical assistance to $5 billion for the 2019–2024 period.

Alba Blue Loan for Recycling in Indonesia. In June 2023, the Asian Development Bank (ADB) signed a $44.2 million nonsovereign blue loan with PT ALBA Tridi Plastics Recycling Indonesia, an ALBA Group Asia company, to establish a polyethylene terephthalate (PET) recycling facility in Central Java.[b] The recycling plant will process PET beverage bottles into high-quality recycled polyethylene terephthalate (rPET) flakes and food-grade rPET pellets, which can be used to produce new rPET bottles. The plant is expected to recycle up to 48,000 tons of PET bottles annually, diverting them from landfills, open burning, or leakage into the ocean. The plant will produce 36,000 tons of rPET, which will offset up to 30,500 tons of carbon dioxide that would have resulted from using virgin PET.

ADB and the Leading Asia's Private Infrastructure Fund will each provide $22.1 million in funding for the project. Blue loans are financing instruments that aim to safeguard access to clean water, protect underwater environments, and invest in a sustainable water economy.

The project supports the Indonesian government's goal of reducing plastic waste leakage by 70% by 2025, and near-zero plastic pollution by 2040. It adheres to ADB's Ocean Finance Framework and the defined criteria for investments under its Action Plan for Healthy Oceans and Sustainable Blue Economies.

[a] ADB. Regional: Indorama Ventures Regional Blue Loan Project. https://www.adb.org/projects/54333-001/main.
[b] ADB. Indonesia: Alba Blue Loan for Recycling. https://www.adb.org/projects/56207-001/main.
Source: ADB based on AWR Lloyd.

ACGF Technical Assistance Support

TA from the ACGF is provided to enhance capacity to support the origination and structuring of green and blue infrastructure projects. The goal is to develop a pipeline of projects that integrate innovative financing mechanisms, which can demonstrate the potential to raise capital at scale. The TA supports upstream studies that identify viable projects, develops and pilots innovative financing mechanisms for early-stage project concepts, and helps structure finance-ready projects.

The Blue SEA Finance Hub

The Blue SEA Finance Hub was launched in November 2021 with TA support from the ACGF. Based in ADB's Indonesia Resident Mission, the hub supports the operationalization of ADB's Action Plan for Healthy Oceans and Sustainable Blue Economies in ASEAN. Specifically, the Blue Hub works with regional governments to identify and structure financially viable ocean health projects, enhance capacities of government stakeholders on blue project frameworks, and promote the upscaling of SMEs.

A cargo ship docking at Danang Port.
The port is the third largest port system in Viet Nam and lies at the eastern end of the Greater Mekong Subregion East–West Economic Corridor, which connects Viet Nam with the Lao People's Democratic Republic, Myanmar, and Thailand.

APPENDIX: KEY PUBLIC SECTOR POLICY ACTIONS, INITIATIVES, AND PROJECTS

Agency	Policy Actions, Initiatives, and Projects
Ministry of Natural Resources and Environment	**POLICY ACTIONS AND INITIATIVES:** **Enhancement and Conservation of the National Environmental Quality Act B.E. 2535**[a] • Designates wetlands as environmentally protected areas regardless of lawful ownership status • Determines environmentally protected areas and pollution control areas and regulates the activities therein **Promotion of Marine and Coastal Resources Management Act B.E. 2558**[b] • Designates marine and coastal protected areas that include mangrove conservation and specific types of marine and coastal resources • Aims to establish laws concerning the management of marine and coastal resources and promote the integration and participation of local communities in conservation efforts **Community Forests Act B.E. 2562**[c] • Aims to promote community participation in the use of natural resources in forests, and in the rehabilitation and conservation of forests, including mangroves and forest areas that are not protected as designated national parks or reserved forests **National Parks Act B.E. 2562**[d] • Aims to protect and preserve designated national parks, which in some cases cover coastal or marine areas and habitat such as mangroves **National Environmental Quality Management Plan (2017–2022)**[e] • Addresses economic implications and mitigative measures for extreme climate events such as floods and droughts • References lack of regulations governing forestry bonds as one of the key reasons proposed bonds were not issued **20-Year National Strategy (2018–2037)**[f] • Outlines strategy to achieve developed country status in accordance with the Sufficiency Economy Philosophy **National Economic and Social Development Plan – The Twelfth Plan (2017–2021)**[g] • Outlines 10 strategies including environmentally friendly, sustainable development **PROJECTS:** **Department of Marine and Coastal Resources (DMCR) – Phuket AquaMUSEUM**[h] • The first Aqua Museum highlighting marine life to promote tourism in Phuket, and for scientific research and educational purposes; project under Her Royal Highness Princess Maha Chakri Sirindhorn Prince Siriwannavari Naree Ratana Rajakanya **DMCR – Dow and Thailand Mangrove Alliance**[i] • The pilot mangrove forestation project will take place in Paknam Prasae subdistrict; intends to set up a carbon credit mechanism, collect data on waste management, and be a "blue carbon" tourist attraction and educational center

Table 2 *continued*

Agency	Policy Actions, Initiatives, and Projects
Ministry of Agriculture and Cooperatives	**POLICY ACTIONS AND INITIATIVES:** **Emergency Decree on Fisheries Act B.E. 2558[j]** • Formally recognizes IUU-fishing under the laws of foreign coastal states and international laws as a crime punishable within Thailand • Reinforces the illegality of unregistered migrant workers or practices against labor protection laws within fisheries operations • Empowers special committee to issue fisheries- and aquaculture-related conservation policies **Land Development Act B.E. 2551[k]** • Authorizes policy prescribing soil and water conservation measures to reduce soil erosion and to prevent landslides • The appointed committee has the duty to specify soil and water conservation areas **2020–2022 National Marine Fisheries Management Plan[l]** • Restore fisheries to support sustainable fishing in deep-sea and overseas waters • Achieve IUU-free fisheries • Create healthy habitats and environments • Improve livelihoods of artisanal fishers and fishing communities • Achieve effective fisheries management capacity • Improve data collection and management for decision-making **PROJECTS:** **Enhancing Climate Resilience in Thailand through Effective Water Management and Sustainable Agriculture[m]** • The project aims to build the resilience of farmers in the Yom and Nan River basins • Fisher supporting the project in the Aquaculture field (โครงการส่งเริมอาชีพประมงจากการเพาะเลี้ยง)[n] • Project objectives include the transfer of knowledge and technology for freshwater aquaculture **Bank of Aquaculture development (โครงการสนับสนุนธนาคารผลผลิตสัตว์น้ำแบบมีส่วนร่วม)[o]** • A marine conservation project that aims to create over 20 development areas to increase aquaculture production across Thailand **Energy efficiency in aquaculture[p]** • A technology development project that supports upscaling and adopting innovations and good practices for energy efficiency in aquaculture
Tourism Authority of Thailand, Ministry of Tourism and Sport	**POLICY ACTIONS AND INITIATIVES:** Outlined a strategy for creating a "new beginning for Thai tourism," emphasizing experience-based and sustainable tourism.[q] **PROJECTS:** **Sustainability development project for marine life within the Eastern Economic Corridor (EEC) zone[r]** • Aims to promote tourism and the restoration of marine resources and create a stable income for people in the EEC **Phuket Health Sandbox (Medical & Wellness hub)[s]** • Aims to connect health and primary care systems through digital technology, develop a Digital Health Platform, be a model for urban health systems; and promote tourism as a "City Connecting People through Global Health."

continued on next page

Table 2 *continued*

Agency	Policy Actions, Initiatives, and Projects
Department of Public Works and Town and Country Planning, Ministry of Interior	POLICY ACTIONS AND INITIATIVES: **Town Planning Act B.E. 2562**[t] • Covers 878 districts across the country • Integrates the water blueprint plan to increase the level of long-term flood prevention • Plan is expected to be presented to the Thai cabinet in 2023 PROJECTS: **Construction of tourism and recreational center (The Park Khao Lak)**[u] • This park project in the southern Phang Nga province is one of six tourism promotion projects in coastal provinces, including Satun, Trang, Phuket, Phang Nga, and Krabi

Bank of Thailand	POLICY ACTIONS AND INITIATIVES:
	
	Consultation Paper on the Financial Landscape (February 2022)[v]
	• Includes transition to the digital economy while coping with environmental risks, building resiliency, and containing systemic risks from unstable environments
	
	Directional Paper on the Financial Landscape (August 2022)[w]
	• Focuses on increasing resilience of the financial sector against environment-related challenges
	• Supports business transitions following the country's net zero targets
	• Five building blocks include:
	  (i) Consideration for environmental factors in FI operations
	  (ii) Establishing a taxonomy that defines and classifies economic activities based on their environmental impact
	  (iii) Developing data platforms for environment-related data and set disclosure standards for FIs to strengthen planned ESG registries, disclosure platforms, data analytics platforms, and product program platforms
	  (iv) Supporting the creation of appropriate incentives, lowering operating costs for businesses, and lowering risk costs for FIs
	  (v) Enhancing knowledge management and skills of financial sector personnel

Securities and Exchange Commission	POLICY ACTIONS AND INITIATIVES: • Implement a mandatory sustainability reporting requirement—"One report"[x] • Investment Governance Code (I-Code): Guideline for responsible and sustainable investment management for institutional investors and issue regulations related to issuance and offer for sale of ESG bonds[y]
Stock Exchange of Thailand (SET)	POLICY ACTIONS AND INITIATIVES: **SET Thailand Sustainability Investment Index (this Index)**[z] • Separate index for sustainability was established to cater to long-term investors **SET Sustainability Reporting guide and ESG metrics**[aa] • Covers ESG materials topics

Thailand Board of Investment	POLICY ACTIONS AND INITIATIVES:
	
	Five-Year Investment Promotion Strategy (2023–2028)[bb]
	• Promotes investments that increase national competitiveness and reduces social and economic disparities
	
	Bio-Circular-Green Economy
	  (i) Promotes bioenergy, biomaterials, biochemicals, and waste–to–energy investments[cc]
	  (ii) Links tourism to the country's capital and promotes tourism in less-visited cities[dd]

continued on next page

Table 2 *continued*

Agency	Policy Actions, Initiatives, and Projects
Ministry of Finance	**POLICY ACTIONS AND INCENTIVES:** **Ministry of Finance and Krung Thai Bank—Thailand Environment Fund**[ee] • Established to issue loans or grants for wastewater treatment or solid waste disposal facilities **Ministry of Finance—Ministry Plans to Issue Bonds**[ff] • Promotes the issuance of blue bonds for the sustainability of ocean and marine resources
Ministry of Transport	**POLICY ACTIONS AND INCENTIVES:** **Civil Liability for Oil Pollution Damage Caused by Ships Act B.E. 2560**[gg] **and Requirement of Contributions to the International Fund for Compensation for Oil Pollution Damage Caused by Ships Act B.E. 2560**[hh] • Formally integrates the two International Conventions on oil pollution damage caused by ships of the International Maritime Organization into domestic law, and together seeks to provide protection and compensation for the damage caused from oil spills within Thailand's territorial waters, which includes environmental damages • Clear allocation of liability to the ship owners prohibiting claims being made against the employees, charterers, operators, etc. **PROJECTS:** **Marine Department - Enlargement of Jomtien Beach**[ii] • Addresses coastal erosion and attracts tourists: widened Bang Saen beach by 50 meters in 2021 and began widening Jomtien Beach in Pattaya in 2022. **Port Authority of Thailand—The Development of Laem Chabang Phase III project**[jj] • Develops Laem Chabang Port as a gateway of trade and investment; this development will handle a container throughput capacity of 4 million TEU/year and is expected to operate by 2025
Ministry of Energy	Department of Alternative Energy Development and Efficiency (DEDE) formerly named "National Energy Authority" is responsible for driving Thailand's transition to renewable energy and reducing the nation's overall energy consumption. DEDE oversees energy promotion, regulation, provision, development, planning, and technology dissemination.[kk]
Office of SMEs Promotion (OSMEP)	August 2022: OSMEP teamed up with the Office of National Higher Education Science Research and Innovation Policy Council (NXPO) to increase innovation capacity and support for SMEs to implement the BCG model and increase market competitiveness. OSMEP and NXPO aims to support 1,000 innovation-driven enterprises and increase revenue by B1 billion ($28.5 million).[ll]

BCG = bio–circular–green; DMCR = Department of Marine and Coastal Resources; ESG = environmental, social, and governance; FI = financial institution;

IUU = illegal, unreported, and unregulated; TEU = twenty-foot equivalent unit.

[a] Government of Thailand, Office of the Council of State. Enhancement and Conservation of the National Environmental Quality Act B.E. 2535. https://www.krisdika.go.th/librarian/get?sysid=396089&ext=pdf.

[b] Government of Thailand, Office of the Council of State. Promotion of Marine and Coastal Resources Management Act B.E. 2558. https://www.krisdika.go.th/librarian/get?sysid=810255&ext=pdf.

[c] Government of Thailand, Office of the Council of State. Community Forests Act B.E. 2562. https://www.krisdika.go.th/librarian/get?sysid=834508&ext=pdf.

[d] Government of Thailand, Office of the Council of State. National Parks Act B.E. 2562. https://www.krisdika.go.th/librarian/get?sysid=834522&ext=pdf.

[e] Government of Thailand, Office of Natural Resources and Environmental Policy and Planning (ONEP). 2017. Thailand Environmental Quality Management Plan 2017–2021. https://data.opendevelopmentmekong.net/dataset/94570d1b-d16c-468b-803e-24ed6efdc920/resource/aba7afd1-91fe-4509-aba8-3b5f9fe5b33c/download/onep-_-neqmp.pdf.

[f] Government of Thailand, Office of the National Economic and Social Development Board, National Strategy Secretariat Office. 2018. National Strategy 2018–2037 (Summary). https://www.bic.moe.go.th/images/stories/pdf/National_Strategy_Summary.pdf.

[g] Government of Thailand, Office of the National Economic and Social Development Board, Office of the Prime Minister. 2017. The Twelfth National Economic and Social Development Plan Summary (2017-2021). https://www.nesdc.go.th/ewt_dl_link.php?nid=9640.

[h] Approved by the Thai Cabinet on 3 November 2020. Government of Thailand, The Secretariat of the Cabinet. 2020. Project Proposals under the Economic and Social Development of the Southern Andaman Provinces "Conservation and Restoration of Marine and Coastal Resources Project for Sustainable Tourism." https://resolution.soc.go.th/?prep_id=402318.

i   Government of Thailand, MNRE, Department of Marine and Coastal Resources. 2020. Dow and Thailand Mangrove Alliance. https://dmcrth.dmcr.go.th/attachment/dw/download.php?WP=nKq4MUNkogy3ZHkCoMOahKGtnJg4WaNlogS3ARj0oH9axUF5nrO4MNo7o3Qo7o3Q.

j   Government of Thailand, Office of the Council of State. Emergency Decree on Fisheries B.E. 2558. https://www.krisdika.go.th/librarian/get?sysid=780563&ext=pdf.

k   Government of Thailand, Office of the Council of State. Land Development Act B.E. 2551. https://www.krisdika.go.th/librarian/get?sysid=809955&ext=pdf.

l   Government of Thailand, Ministry of Agriculture and Cooperatives, Department of Fisheries. 2020. Marine Fisheries Management Plan of Thailand 2020-2022. https://faolex.fao.org/docs/pdf/tha212512.pdf.

m   United Nations Development Programme, Climate Change Adaptation. 2021. Enhancing Climate Resilience in Thailand through Effective Water Management and Sustainable Agriculture Project. https://www.adaptation-undp.org/projects/enhancing-climate-resilience-thailand-through-effective-water-management-and-sustainable.

n   Government of Thailand, Ministry of Agriculture and Cooperatives, Department of Fisheries. 2022. Fishery Career Promotion Project. https://www4.fisheries.go.th/local/file_document/20220809094806_1_file.pdf.

o   Government of Thailand, Ministry of Agriculture and Cooperatives, Department of Fisheries. 2022. Bank of Aquaculture Development. https://www4.fisheries.go.th/local/pic_activities/202006111445241_pic.pdf.

p   Government of Thailand, Ministry of Agriculture and Cooperatives, Department of Fisheries,. 2022. Innovation and Good Practices to Increase Energy Efficiency in Aquaculture. https://www4.fisheries.go.th/dof/news_local/1210/149199.

q   Bangkok Post. 2022. Thailand Revitalising Tourism Locally and Internationally for Asia-Pacific. 23 October. https://www.bangkokpost.com/business/2419736/thailand-revitalising-tourism-locally-and-internationally-for-asia-pacific.

r   Eastern Economic Corridor. 2022. Tourists for Sustainable Development: Sea Creature. https://www.eeco.or.th/th/news/443.

s   The Phuket News. 2021. Phuket Health Sandbox Approved with B85mn Budget. 17 November. https://www.thephuketnews.com/phuket-health-sandbox-approved-with-b85mn-budget-82064.php.

t   Government of Thailand, Department of Public Works and Town & Country. 2019. Town Planning Act B.E.2562. http://subsites.dpt.go.th/edocument/images/pdf/doc_urban/11.pdf.

u   The Nation Thailand. 2022. Cabinet Okays THB338m for Tourism Projects in 5 Andaman Coast Provinces. 15 June. https://www.nationthailand.com/in-focus/40016651.

v   Bank of Thailand. 2022. Repositioning Thailand's Financial Sector for a Sustainable Digital Economy. https://www.bot.or.th/content/dam/bot/financial-innovation/financial-landscape/ConsultationPaper-FinancialLandscape-EN.pdf.html.

w   Bank of Thailand. 2022a. Directional Paper on Transitioning Towards Environmental Sustainability under the New Thai Financial Landscape. BOT Press Release No. 43/2022. 23 August. https://www.bot.or.th/en/news-and-media/news/news-20220823.html.

x   Government of Thailand, Securities and Exchange Commission. 2020. Manual for Preparing the Annual Registration Statement and Annual Report: 56-1 One Report. https://publish.sec.or.th/nrs/8619s.pdf.

y   Government of Thailand, Securities and Exchange Commission. 2022. Investment Governance Code for Institutional Investors (I Code). https://www.sec.or.th/cgthailand/EN/Documents/ICode/ICodeBookEN.pdf.

z   Stock Exchange of Thailand. 2022. Thailand Sustainability Investment (THSI). https://www.setsustainability.com/page/thsi-thailand-sustainability-investment.

aa   Stock Exchange of Thailand. 2018. SET Sustainability Reporting Guidelines. https://www.setsustainability.com/download/yeu6plait5f4gb7.

bb   The Board of Investment of Thailand. 2023. Investment Promotion Guide 2023. https://www.boi.go.th/upload/content/BOI_A_Guide_EN.pdf.

cc   C. Kaewsang. 2020. Opportunities in the Bio-Circular-Green (BCG) Economy and BOI Support Measures. The Board of Investment of Thailand. https://www.boi.go.th/upload/content/BOI-BCG_DSG%20Chokedee.pdf.

dd   Bangkok Post. 2021. Reimagining Thailand with a BCG Economy. 10 September. https://www.bangkokpost.com/business/2179811/reimagining-thailand-with-a-bcg-economy.

ee   K. Choeypun. Thailand Environment Fund. Convention on Biological Diversity. https://www.cbd.int/financial/trustfunds/Thailand-ef.pdf.

ff   Royal Thai Embassy, Washington, DC. 2021. Finance Ministry Plans to Issue "Blue Bonds." https://thaiembdc.org/2021/04/21/finance-ministry-plans-to-issue-blue-bonds/.

gg   Government of Thailand, Office of the Council of State. Civil Liability for Oil Pollution Damage Caused by Ships Act B.E. 2560. https://www.krisdika.go.th/librarian/get?sysid=809776&ext=pdf.

hh   Government of Thailand, Office of the Council of State. Requirement of Contributions to the International Fund for Compensation for Oil Pollution Damage Caused by Ships Act B.E. 2560. http://web.krisdika.go.th/data/document/ext810/810219_0001.pdf.

ii   C. Pupattanapong. 2020. Work to Start on B586m Enlargement of Jomtien Beach. Bangkok Post. 12 June. https://www.bangkokpost.com/thailand/general/1933848/work-to-start-on-b586m-enlargement-of-jomtien-beach.

jj   Port Technology International. 2021. Gulf Energy-led Consortium Signs $927 million Deal on Laem Chabang Phase III Project. 2 December. https://www.porttechnology.org/news/gulf-energy-led-consortium-signs-927-million-deal-on-laem-chabang-phase-iii-project/.

kk   Government of Thailand, Ministry of Energy, Department of Alternative Energy Development and Efficiency. 2022. History of DEDE. https://weben.dede.go.th/webmax/content/history-dede.

ll   Government of Thailand, Ministry of Higher Education, Science, Research and Innovation, Office of National Higher Education, Science, Research, and Innovation Policy Council. 2022. NXPO and OSMEP Team Up to Strengthen Innovation Capacity of Thai SMEs. 23 August. https://www.nxpo.or.th/th/en/12575/.

Source: AWR Lloyd.

REFERENCES

Apisitniran, L. 2019. Treatment Touted for 10% Growth. *Bangkok Post.* 11 March. https://www.bangkokpost.com/business/1642368/treatment-touted-for-10-growth.

Asian Development Bank (ADB). Healthy Oceans and Sustainable Blue Economies. https://www.adb.org/what-we-do/topics/environment/healthy-oceans-sustainable-blue-economies.

ADB. Indonesia : Alba Blue Loan for Recycling. https://www.adb.org/projects/56207-001/main.

———. Regional: Indorama Ventures Regional Blue Loan Project. https://www.adb.org/projects/54333-001/main.

———. 2022a. ADB Approves $3.8 Million Support for Development of Coral Reef Insurance News Release. https://www.adb.org/news/adb- approves-3-8-million-support-development-coral-reef-insurance.

———. 2022b. *Asian Development Bank Ocean Finance Framework.* Manila. https://www.adb.org/publications/adb-ocean-finance-framework.

———. 2022c. *Financing the Blue Economy: Investments in Sustainable Blue Small-Medium Enterprises and Projects in Asia and the Pacific.* Manila. https://www.adb.org/sites/default/files/publication/806136/financing-blue-economy.pdf.

ADB. 2021. *Financing the Ocean Back to Health in Southeast Asia.* Manila. https://www.adb.org/sites/default/files/publication/756686/financing-ocean-health-southeast-asia.pdf.

Asian Development Blog. 2019. *Thailand in Need of 'Energy 4.0'.* https://blogs.adb.org/blog/thailand-need-energy-40.

Bangkok Post. 2021. Reimagining Thailand with a BCG Economy. 10 September. https://www.bangkokpost.com/business/2179811/reimagining-thailand-with-a-bcg-economy.

———. 2022a. Thailand Revitalising Tourism Locally and Internationally for Asia-Pacific. 23 October. https://www.bangkokpost.com/business/2419736/thailand-revitalising-tourism-locally-and-internationally-for-asia-pacific.

———. 2022b. Shrimp Industry Continues to Tread Water. 15 December. https://www.bangkokpost.com/business/2460805/shrimp-industry-continues-to-tread-water.

———. 2023. 3-Airport High-speed Rail Link Completion Seen by 2029. 29 January. https://www.bangkokpost.com/thailand/general/2493950/3-airport-high-speed-rail-link-completion-seen-by-2029.

Bank of Thailand (BOT). Financial Landscape: Environmental Sustainability. https://app.bot.or.th/landscape/en/paper/sustainable/environment/.

———. 2022a. Directional Paper on Transitioning Towards Environmental Sustainability under the New Thai Financial Landscape. BOT Press Release No. 43/2022. 23 August. https://www.bot.or.th/en/news-and-media/news/news-20220823.html.

———. 2022b. TBA Launches ESG Declaration, a Strong Collective Commitment to Expediting Sustainable Development toward Better and Greener Economy. 29 August. https://www.bot.or.th/en/news-and-media/news/news-20220829-2.html.

———. 2022c. Repositioning Thailand's Financial Sector for a Sustainable Digital Economy. https://www.bot.or.th/content/dam/bot/financial-innovation/financial-landscape/ConsultationPaper-FinancialLandscape-EN.pdf.html.

Barbier, E. B. 2006. Mangrove Dependency and the Livelihoods of Coastal Communities in Thailand. In C. T. Hoanh, T. P. Tuong, J. W. Gowing, and B. Hardy, eds. Environment and Livelihoods in Tropical Coastal Zones: Managing Agriculture-Fishery-Aquaculture Conflicts. http://dx.doi.org/10.1079/9781845931070.0126.

The Biodiversity Finance Initiative. 2018. Thailand. https://www.biofin.org/thailand.

Bloomberg NEF. 2022. Sustainable Debt Issuance Breezed past $1.6 Trillion in 2021. 12 January. https://about.bnef.com/blog/sustainable-debt-issuance-breezed-past-1-6-trillion-in-2021/.

The Board of Investment of Thailand (BOI Thailand). 2020a. Opportunities in the Bio-Circular-Green (BCG) Economy and BOI Support Measures. https://www.boi.go.th/upload/content/BOI-BCG_DSG%20Chokedee.pdf.

———. 2020b. Thailand's Electrical Market. https://www.boi.go.th/index.php?page=electricity.

———. 2021. Seaports. https://www.boi.go.th/index.php?page=seaports.

———. 2023. Investment Promotion Guide 2023. https://www.boi.go.th/upload/content/BOI_A_Guide_EN.pdf.

Boocharoen, C. and A. K. Anal. 2021. Attitudes, Perceptions, and On-farm Self-reported Practices of Shrimp Farmers towards Adoption of Good Aquaculture Practices (GAP) in Thailand. Sustainability. 13 (9). 5194. https://doi.org/10.3390/su13095194.

BTS Group. 2021. BTS Group Jointly Establishes the Carbon Markets Club, Promoting Carbon Credit Trading to Help Reduce Greenhouse Gases Towards a Net Zero Society. 29 June. https://www.btsgroup.co.th/en/update/news-event/688/bts-group-jointly-establishes-the-carbon-markets-club-promoting-carbon-credit-trading-to-help-reduce-greenhouse-gases-towards-a-net-zero-society.

Bunjongsiri, K. 2019. The Overview of Carbon Credit Market in Thailand. SAU Journal of Science & Technology. 5 (2). pp. 1–9. https://ph01.tci-thaijo.org/index.php/saujournalst/article/view/184556.

California Ocean Protection Council. 2018. California Fisheries Fund. 7 September. https://www.opc.ca.gov/2010/01/california-fisheries-fund/.

Carbon Offset Guide. Mandatory & Voluntary Offset Markets. https://www.offsetguide.org/understanding-carbon-offsets/carbon-offset-programs/mandatory-voluntary-offset-markets/.

Chakrabongse, D. 2021. Thailand's Subsidies Are Now the Biggest Threat to its Fisheries. Thai Enquirer. 8 April. https://www.thaienquirer.com/26235/thailands-subsidies-are-now-the-biggest-threat-to-its-fisheries/.

Charoen Pokphand Foods (CPF). 2017. CPF Reduces Fishmeal in Aquatic Feed Production, to Promote Sustainable Use of Marine Resources. 21 July. https://www.cpfworldwide.com/en/media-center/1034.

Choeypun, K. *Thailand Environment Fund.* Convention on Biological Diversity. https://www.cbd.int/financial/trustfunds/Thailand-ef.pdf.

Circulate Capital. Investments. https://www.circulatecapital.com/investments.

Climate Policy Initiative. 2022. *Indonesia Blue Finance Landscape.* https://www.climatepolicyinitiative.org/id/publication/indonesia-blue-finance-landscape/.

Cooperative League of Thailand. Fishing Cooperatives. https://www.cltcoop.com/17624383/fishery-cooperatives.

Deutsche Welle. 2018. Thailand's Tourism Boom Damages Corals. 30 January. https://www.dw.com/en/thailands-tourism-boom-damages-corals-to-critical-level/a-42361812.

Eastern Economic Corridor. 2022. Tourists for Sustainable Development: Sea Creature. https://www.eeco.or.th/th/news/443.

Electricity Generating Authority of Thailand. 2021. The World's Largest Hydro-floating Solar Hybrid. 1 July. https://www.egat.co.th/home/en/the-worlds-largest-hydro-floating-solar-hybrid/.

Enviliance Asia. 2022. *Another Step Toward PRTR Implementation in Thailand.* https://enviliance.com/regions/southeast-asia/th/th-chemical/th-prtr.

Frias, J. A. U., and R. Kumar. 2022. *Creating Markets in Thailand: Rebooting Productivity for Resilient Growth.* Country Private Sector Diagnostic (CPSD) Washington, DC: World Bank Group. https://documents.worldbank.org/en/publication/documents-reports/documentdetail/468721645451588595/creating-markets-in-thailand-rebooting-productivity-for-resilient-growth.

Government of Thailand, Department of Public Works and Town & Country. 2019. *Town Planning Act B.E.2562.* http://subsites.dpt.go.th/edocument/images/pdf/doc_urban/11.pdf.

Government of Thailand, Energy Policy and Planning Office. 2018. *Thailand's Power Development Plan (PDP) 2018-2037 (TH).* https://policy.thinkbluedata.com/sites/default/files/Thailand%E2%80%99s%20Power%20Development%20Plan%20%28PDP%29%20%282018%E2%80%932037%29%20%28TH%29.pdf.

Government of Thailand, Ministry of Agriculture and Cooperatives, Department of Fisheries. 2019. *Marine Fisheries Management Plan of Thailand 2020–2022.* https://faolex.fao.org/docs/pdf/tha212512.pdf.

————. 2020b. *Marine Fisheries Management Plan of Thailand 2020-2022.* https://www4.fisheries.go.th/local/file_document/20220912132213_1_file.pdf.

————. 2022a. *Bank of Aquaculture Development.* https://www4.fisheries.go.th/local/pic_activities/202006111445241_pic.pdf.

————. 2022b. *Fishery Career Promotion Project.* https://www4.fisheries.go.th/local/file_document/20220809094806_1_file.pdf.

————. 2022c. *Innovation and Good Practices to Increase Energy Efficiency in Aquaculture.* https://www4.fisheries.go.th/dof/news_local/1210/149199.

Government of Thailand, Ministry of Agriculture and Cooperatives, Cooperative Promotion Department. 2008. *Fishing Cooperatives.* https://www.cpd.go.th/cpden/images/FisheriesCooperatives.pdf.

Government of Thailand, Ministry of Agriculture and Cooperatives, National Bureau of Agricultural Commodity and Food Standards. 2014. *Good Aquaculture Practices for Marine Shrimp Farm.* https://www.acfs.go.th/standard/download/eng/GAP-FOR-MARINE-SHRIMP-FARM_EN.pdf.

Government of Thailand, Ministry of Energy. 2018. *Thailand's Power Development Plan 2018-2037 (Rev. 1).* https://policy.asiapacificenergy.org/sites/default/files/Thailand's%20Power%20Development%20Plan%20%28PDP%29%20%282018–2037%29%20%28TH%29.pdf.

Government of Thailand, Ministry of Energy, Department of Alternative Energy Development and Efficiency. 2018. *Final Report on the Status of Electricity Generation from Solar in Thailand.* PDF_PVstatus2561.pdf (dede.go.th).

———. 2022. History of DEDE. https://weben.dede.go.th/webmax/content/history-dede.

Government of Thailand, Ministry of Higher Education, Science, Research and Innovation, Office of National Higher Education, Science, Research, and Innovation Policy Council. 2022. NXPO and OSMEP Team Up to Strengthen Innovation Capacity of Thai SMEs. 23 August. https://www.nxpo.or.th/th/en/12575/.

Government of Thailand, Ministry of Natural Resources and Environment, Office of the Natural Resources and Environmental Policy and Planning. 2017. *The National Environment Quality Management Plan 2017–2021.* https://www.onep.go.th/ebook/spd/environment-plan-2560-2564.pdf.

Government of Thailand, Ministry of Natural Resources and Environment, Department of Marine and Coastal Resources. 2020. *Dow and Thailand Mangrove Alliance.* https://dmcrth.dmcr.go.th/attachment/dw/download.php?WP=nKq4MUNkogy3ZHkCoMOahKGtnJg4WaNlogS3ARj0oH9axUF5nrO4MNo7o3Qo7o3Q.

———. 2022. *Guidelines for Utilization of Marine and Coastal Resources under the Blue Economy Framework in Trat Province.* https://www.dmcr.go.th/detailLib/6384.

Government of Thailand, Office of the Council of State. *Civil Liability for Oil Pollution Damage Caused by Ships Act B.E. 2560.* https://www.krisdika.go.th/librarian/get?sysid=809776&ext=pdf.

———. *Community Forests Act B.E. 2562.* https://www.krisdika.go.th/librarian/get?sysid=834508&ext=pdf.

———. *Emergency Decree on Fisheries B.E. 2558.* https://www.krisdika.go.th/librarian/get?sysid=780563&ext=pdf.

———. *Enhancement and Conservation of the National Environmental Quality Act B.E. 2535.* https://www.krisdika.go.th/librarian/get?sysid=396089&ext=pdf.

Government of Thailand, Office of the Council of State. *Land Development Act B.E. 2551.* https://www.krisdika.go.th/librarian/get?sysid=809955&ext=pdf.

———. *Promotion of Marine and Coastal Resources Management Act B.E. 2558.* https://www.krisdika.go.th/librarian/get?sysid=810255&ext=pdf.

———. *Requirement of Contributions to the International Fund for Compensation for Oil Pollution Damage Caused by Ships Act B.E. 2560.* http://web.krisdika.go.th/data/document/ext810/810219_0001.pdf.

Government of Thailand, Office of the National Economic and Social Development Board, National Strategy Secretariat Office. 2018. *National Strategy 2018–2037 (Summary)*. https://www.bic.moe.go.th/images/stories/pdf/National_Strategy_Summary.pdf.

Government of Thailand, Office of the National Economic and Social Development Board, Office of the Prime Minister. 2017. *The Twelfth National Economic and Social Development Plan Summary (2017-2021)*. https://www.nesdc.go.th/ewt_dl_link.php?nid=9640.

Government of Thailand, Office of the National Economic and Social Development Council. 2020. *Thailand's Logistics Report 2020*. https://www.nesdc.go.th/ewt_dl_link.php?nid=11975.

Government of Thailand, Office of Natural Resources and Environmental Policy and Planning. 2017. Thailand Environmental Quality Management Plan 2017–2021. https://data.opendevelopmentmekong.net/dataset/94570d1b-d16c-468b-803e-24ed6efdc920/resource/aba7afd1-91fe-4509-aba8-3b5f9fe5b33c/download/onep-_-neqmp.pdf.

Government of Thailand, The Secretariat of the Cabinet. 2020. Project Proposals under the Economic and Social Development of the Southern Andaman Provinces "Conservation and Restoration of Marine and Coastal Resources Project for Sustainable Tourism." https://resolution.soc.go.th/?prep_id=402318.

Government of Thailand, Securities and Exchange Commission. 2020. *Manual for Preparing the Annual Registration Statement and Annual Report: 56-1 One Report*. https://publish.sec.or.th/nrs/8619s.pdf.

————. 2021. *Strategic Plan 2020–2023*. p. 29. https://www.sec.or.th/TH/Documents/strategicplan/strategicplan-2564-2566.pdf.

————. 2022. *Investment Governance Code for Institutional Investors (I Code)*. https://www.sec.or.th/cgthailand/EN/Documents/ICode/ICodeBookEN.pdf.

Government of the United States, Department of Agriculture, Foreign Agriculture Service, Global Agricultural Information Network. 2018. *Thailand Seafood Report*. https://apps.fas.usda.gov/newgainapi/api/report/downloadreportbyfilename?filename=Seafood%20Report_Bangkok_Thailand_5-8-2018.pdf.

Green Climate Fund. 2020. *Increasing Resilience to Climate Change Impacts in Marine and Coastal Areas along the Gulf of Thailand*. https://www.greenclimate.fund/sites/default/files/document/tha-rs-006.pdf.

Green Finance Institute. Althelia Sustainable Ocean Fund. https://www.greenfinanceinstitute.co.uk/gfihive/case-studies/sustainable-ocean-fund/.

Greenpeace. 2018. *Renewable Energy Job Creation in Thailand*. https://www.greenpeace.or.th/report/Renewable-Energy-Job-Creation-in-Thailand-EN.pdf.

————. 2021. *The Projected Economic Impact of Extreme Sea-Level Rise in Seven Asian Cities in 2030*. https://www.greenpeace.org/static/planet4-eastasia-stateless/2021/06/966e1865-gpea-asian-cites-sea-level-rise-report-200621-f-3.pdf.

Hunton Andrews Kurth. 2022. Hunton Andrews Kurth Secures First Maritime Sustainability Bond in Thailand for TMB Thanachart. 2 June. https://www.huntonak.com/en/news/hunton-secures-first-maritime-sustainability-bond-in-thailand-for-tmb-thanachart-with-intl-finance-corporation-as-the-subscriber.html.

Indorama Ventures. 2021. Indorama Ventures Issues Thb 10 Billion Sustainability-Linked Bond Driving Climate Action and Sustainable Production. 3 November. https://www.indoramaventures.com/en/updates/other-release/1840/indorama-ventures-issues-thb-10-billion-sustainability-linked-bond-driving-climate-action-and-sustainable-production.

Kaewsang, C. 2020. *Opportunities in the Bio-Circular-Green (BCG) Economy and BOI Support Measures.* The Board of Investment of Thailand. https://www.boi.go.th/upload/content/BOI-BCG_DSG%20Chokedee.pdf.

Kasikorn Bank. 2021. *KBank's TCFD Report 2021: Task Force on Climate-related Financial Disclosures.* https://www.kasikornbank.com/SiteCollectionDocuments/sustainable-development/pdf/kbank-tcfd-report2021-th.pdf.

Kingdom of the Netherlands, Open Development Mekong. 2016. *The Water Sector in Thailand.* https://data.opendevelopmentmekong.net/dataset/379173fa-bc66-4866-87ca-00cc73e8139f/resource/b9547f82-089d-438c-91e8-8cc87d860dc4/download/factsheet-the-water-sector-in-thailand-3.pdf.

Kingdom of Thailand. 2021. *Thailand's Overall Maritime Strategy (unofficial translation).* https://md.go.th/wp-content/uploads/2021/03/Thailands-Overall-Maritime-Strategy.pdf.

Mahidol University, Faculty of Environment and Resource Stuides. 2022. *Thai Coast: Vulnerability, Resilience, and Adaptation in Thailand.* https://en.mahidol.ac.th/index.php/sdgs/goal-13-climate-action/94-sdg/2565/2022-10-04-06-16-55/746-thai-coast-coastal-vulnerability-resilience-and-adaptation-in-thailand.

Marks, D., M. A. Miller, and S. Vassanadumrongdee. 2020. The Geopolitical Economy of Thailand's Marine Plastic Pollution Crisis. *Asia Pacific Viewpoint.* 61 (2). pp. 266–282. https://onlinelibrary.wiley.com/doi/full/10.1111/apv.12255.

Modi, A. and M. Lackovic. 2021. Investment and Innovation in Thai Renewable Energy. *Bangkok Post.* 15 March. https://www.bangkokpost.com/business/2083795/investment-and-innovation-in-thai-renewable-energy.

The Nation Thailand. 2017. SCB Expands SME Market with 150% LTV Lending for up to 30 Years. 26 January. https://www.nationthailand.com/breaking-news/30305006.

———. 2022a. Recurring Disease has Taken a Heavy Toll on Thai Shrimp Yields. 16 February. https://www.nationthailand.com/blogs/business/40012418.

———. 2022b. Port Authority Hoists Revenue to Bt15.6 Billion in 2021. 4 March. https://www.nationthailand.com/business/40013020.

———. 2022c. Cabinet Okays THB338m for Tourism Projects in 5 Andaman Coast Provinces. 15 June. https://www.nationthailand.com/in-focus/40016651.

———. 2022d. Thailand to Import 10,500 Tonnes of Shrimp as Domestic Yield Sinks. 8 August. https://www.nationthailand.com/business/40018647.

———. 2023. *Carbon Credit Trading Opens on FTIX Platform Tomorrow.* 15 January. https://www.nationthailand.com/thailand/economy/40024038.

Open Development Mekong, Thailand. 2017. Environment and Natural Resources. 19 December. https://thailand.opendevelopmentmekong.net/topics/environment-and-natural-resources/.

Organisation for Economic Co-operation (OECD). 2021. The Role of Guarantees in Blended Finance. *OECD Development Cooperation Working Papers No. 97.* https://www.oecd-ilibrary.org/docserver/730e1498-en.pdf?expires=1669175087&id=id&accname=guest&checksum=E1E0557F264D2536E1012FE279550A76.

————. 2022. *Financing SMEs and Entrepreneurs 2022: An OECD Scoreboard: Thailand.* https://www.oecd-ilibrary.org/sites/b854dc2c-en/index.html?itemId=/content/component/b854dc2c-en.

Partnerships in Environmental Management for the Seas of East Asia. 2019. *ICM Solutions, Gateway to a Blue Economy: Port Safety, Health, and Environmental Management in the Port Authority of Thailand— Bangkok and Laem Chabang ports.* http://pemsea.org/sites/default/files/KP%2024_0_0.pdf.

Partnerships in Environmental Management for the Seas of East Asia, Department of Marine and Coastal Resources. 2019. *National State of Oceans and Coasts 2018: Blue Economy Growth of Thailand.* https://seaknowledgebank.net/sites/default/files/NSOC%20Thailand%202018%20%28FINAL%29%2012032020.pdf.

The Phuket News. 2021. *Phuket Health Sandbox Approved with B85mn Budget.* 17 November. https://www.thephuketnews.com/phuket-health-sandbox-approved-with-b85mn-budget-82064.php.

Port Technology International. 2021. Gulf Energy-Led Consortium Signs $927 Million Deal on Laem Chabang Phase III Project. 2 December. https://www.porttechnology.org/news/gulf-energy-led-consortium-signs-927-million-deal-on-laem-chabang-phase-iii-project/.

Priyashnatha, A. K. H., and U. Edirisinghe. 2021. Lessons Learnt from the Past to Mitigate the Negative Aspects of Aquaculture in Developing Countries. *PSAKU International Journal of Interdisciplinary Research.* 10 (2). https://papers.ssrn.com/sol3/papers.cfm?abstract_id=3964762.

Pupattanapong, C. 2020. Work to Start on B586m Enlargement of Jomtien Beach. *Bangkok Post.* 12 June. https://www.bangkokpost.com/thailand/general/1933848/work-to-start-on-b586m-enlargement-of-jomtien-beach.

Ranthodsang, M. et al. 2020. Offshore Wind Power Assessment on the Western Coast of Thailand. *Energy Reports.* 6. pp. 1135–1146.

Rolt, A. 2022. The Nature Conservancy Debuts New Hawaii Coral Reef Insurance Plan. *Green Biz.* https://www.greenbiz.com/article/nature-conservancy-debuts-new-hawaii-coral-reef-insurance-plan.

Royal Thai Embassy, Washington, DC. 2021. *Finance Ministry Plans to Issue "Blue Bonds."* https://thaiembdc.org/2021/04/21/finance-ministry-plans-to-issue-blue-bonds/.

Rudjanakanoknad, J., W. Suksirivoraboot, and S. Sukdanont. 2014. Evaluation of International Ports in Thailand through Trade Facilitation Indices from Freight Forwarders. *Procedia – Social and Behavioral Sciences.* 111 (2014). pp. 1073–1082. https://www.sciencedirect.com/science/article/pii/S1877042814001438.

Sampantamit, T. et al. 2020. Aquaculture Production and Its Environmental Sustainability in Thailand: Challenges and Potential Solutions. *Sustainability.* 12 (5). https://doi.org/10.3390/su12052010.

Sawasklin, P., S. Saeung, and J. Taweekun. 2021. Study on Offshore Wind Energy Potential in the Gulf of Thailand. *International Journal of Renewable Energy Research.* 11 (4). pp. 1947–1958. https://www.ijrer.org/ijrer/index.php/ijrer/article/view/12213/pdf.

Semmahasak, S. 2014. *Soil Erosion and Sediment Yield in Tropical Mountainous Watershed of Northwest Thailand: The Spatial Risk Assessments Under Land Use and Rainfall Changes.* School of Geography, Earth and Environmental Sciences, College of Life and Environmental Sciences, University of Birmingham. https://core.ac.uk/reader/33527420; NBSAP Forum. 2018. Combating Industrial Pollution in Thailand. https://nbsapforum.net/knowledge-base/best-practice/combating-industrial-pollution-thailand.

Statista. 2022a. Water Quality of Surface Water Surfaces in Thailand 2021. https://www.statista.com/statistics/1295279/thailand-condition-of-surface-water-sources/(accessed February 2023).

———. 2022b. Tourism Industry in Thailand – Statistics & Facts, International and Domestic Tourists' Revenue. https://www.statista.com/topics/6845/tourism-industry-in-thailand/#topicHeader__wrapper (accessed February 2023).

Stock Exchange of Thailand. 2018. *SET Sustainability Reporting Guidelines.* https://www.setsustainability.com/download/yeu6plait5f4gb7.

———. 2022. *Thailand Sustainability Investment (THSI).* https://www.setsustainability.com/page/thsi-thailand-sustainability-investment.

Sukphisit, S. 2019. Troubled Waters. *Bangkok Post.* 28 July. https://www.bangkokpost.com/life/social-and-lifestyle/1720327/troubled-waters.

Suwannapoom, S. 2020. County Fisheries Trade: Thailand. *Southeast Asian Fisheries Development Center (SEAFDEC).* 2 April. http://www.seafdec.org/county-fisheries-trade-thailand/.

Tanakasempipat, P. 2020. Plastic Piles Up in Thailand as Pandemic Efforts Sideline Pollution Fight. *Reuters.* 11 May. https://www.reuters.com/article/us-health-coronavirus-thailand-plastic-idUSKBN22N12W.

Task Force on Climate-related Financial Disclosure. 2017. *Recommendations of the Task Force on Climate-related Financial Disclosures.* https://assets.bbhub.io/company/sites/60/2020/10/FINAL-2017-TCFD-Report-11052018.pdf.

Thai Credit Guarantee Corporation. 2022. *Annual Report 2021.* https://www.tcg.or.th/uploads/file/221010103026gRHE.pdf.

Thai Union. 2021a. Thai Union Launches Inaugural Sustainability-linked Loan. 16 February. https://www.thaiunion.com/en/newsroom/press-release/1292/thai-union-launches-inaugural-sustainability-linked-loan.

———. 2021b. Thai Union Feedmill Aims to Lead Aquaculture and Commercial Feed Sector, Following Listing on Stock Exchange of Thailand. 29 October. https://www.thaiunion.com/en/thai-union-cares/leadership/1475/thai-union-feedmill-aims-to-lead-aquaculture-and-commercial-animal-feed-sector-following-listing-on-stock-exchange-of-thailand.

———. 2022. Healthy Living Healthy Oceans. 56-1 *One Report.* https://investor.thaiunion.com/misc/ar/20220325-tu-or2022-en.pdf.

Thai–German Cooperation. 2018. ECAM Tool Gears Up Water Sector Towards Achieving Greenhouse Gas Reduction Target. 20 April. https://www.thai-german-cooperation.info/en_US/ecam-tool-gears-up-water-sector-towards-achieving-greenhouse-gas-reduction-target/.

Thailand Development Research Institute. 2019. *Tackling Thailand's Food-waste Crisis*. https://tdri.or.th/en/2019/10/tackling-thailands-food-waste-crisis/.

Thailand Environment Institute. 2021. *Waste to Energy*. https://www.tei.or.th/en/article_detail.php?bid=49.

Thansettakij. 2022. Mangrove Planting to Reduce Carbon Credits. 11 May. https://www.thansettakij.com/business/524532.

The Nature Conservancy. 2022. The Nature Conservancy Announces First-ever Coral Reef Insurance Policy in the US. 21 November. https://www.nature.org/en-us/newsroom/first-ever-us-coral-reef-insurance-policy/.

Theparat, C. 2022. Panel Okays B1.92bn Rice Insurance Plan. *Bangkok Post.* 28 April. https://www.bangkokpost.com/business/2301478/panel-okays-b1-92bn-rice-insurance-plan.

TMBThanachart Bank Public Company Limited. A Fact Sheet on Business Credits with Collateral. https://media.ttbbank.com/1/document/sbo/sales_sheet_so_smooth.pdf.

———. 2022. *TTB Green and Blue Bond Framework.* https://media.ttbbank.com/1/ir/green-blue-bond/ttb-green-blue-bond-framework-en.pdf.

Toomgum, S. 2022. Thais Urged to Embrace Disruptive Tech. *Bangkok Post.* 23 February. https://www.bangkokpost.com/business/2268483/thais-urged-to-embrace-disruptive-tech.

TTG Asia. 2018. Thai Sustainable Tourism Standard Gets GSTC Recognition. 11 May. https://www.ttgasia.com/2018/05/11/thai-sustainable-tourism-standard-gets-gstc-recognition/221195/.

United Nations Department of Economic and Social Affairs. 2016. *Thailand towards Sustainable Management of Marine and Coastal Habitats.* https://sdgs.un.org/partnerships/thailand-towards-sustainable-management-marine-and-coastal-habitats.

United Nations Development Programme. 2022. What Are Carbon Markets and Why Are They Important? 18 May. https://climatepromise.undp.org/news-and-stories/what-are-carbon-markets-and-why-are-they-important#:~:text=In%20a%20nutshell%2C%20carbon%20markets,gas%20reduced%2C%20sequestered%20or%20avoided.

United Nations Development Programme, Climate Change Adaptation. 2021. Enhancing Climate Resilience in Thailand through Effective Water Management and Sustainable Agriculture Project. https://www.adaptation-undp.org/projects/enhancing-climate-resilience-thailand-through-effective-water-management-and-sustainable.

United Nations Environment Programme. Sustainable Blue Economy. The Principles: Sustainable Blue Finance. https://www.unepfi.org/blue-finance/the-principles/.

United Nations Framework Convention on Climate Change. 2021. *Mid-century, Long-term Low Greenhouse Gas Emissions Development Strategy, Thailand.* https://unfccc.int/sites/default/files/resource/Thailand_LTS1.pdf.

Verra. 2020. First Blue Carbon Conservation Methodology Expected To Scale Up Finance For Coastal Restoration and Conservation Activities. 9 September. https://verra.org/first-blue-carbon-conservation-methodology-expected-to-scale-up-finance-for-coastal-restoration-conservation-activities/.

———. 2021. Verra Has Registered Its First Blue Carbon Conservation Project. 10 May. https://verra.org/press-release-verra-has-registered-its-first-blue-carbon-conservation-project/.

Waewsak, J., M. Landry, and Y. Gagnon. 2015. Offshore Wind Power Potential of the Gulf of Thailand. *Renewable Energy.* 81 (C). pp. 609–626. https://www.sciencedirect.com/science/article/abs/pii/S0960148115002517.

Wipatayotin, A. 2021. New Coastal Checklist Gets the Nod. *Bangkok Post.* 28 September. https://www.bangkokpost.com/thailand/general/2188571/new-coastal-checklist-gets-the-nod.

Working Group on Sustainable Finance. 2021. *Sustainable Finance Initiatives for Thailand.* https://www.sec.or.th/TH/Documents/KnowledgeBase/SustainableFinanceInitiativesforThailand.pdf.

World Bank. 2017. What is the Blue Economy? 6 June. https://www.worldbank.org/en/news/infographic/2017/06/06/blue-economy.

———. 2022. *Plastic Waste Material Flow Analysis for Thailand.* https://documents1.worldbank.org/curated/en/099515103152238081/pdf/P17099409744b50fc09e7208a58cb52ae8a.pdf.

World Bank and Asian Development Bank. 2021. *Thailand Climate Risk Country Profile.* https://climateknowledgeportal.worldbank.org/sites/default/files/2021-08/15853-WB_Thailand%20Country%20Profile-WEB_0.pdf.

World Shipping Council. *The Top 50 Container Ports.* 2020. https://www.worldshipping.org/top-50-ports.

Yenpoeng, T. 2017. Fisheries Country Profile: Thailand. SEAFDEC. http://www.seafdec.org/fisheries-country-profile-thailand/.

A Road Map for Scaling Up Private Sector Financing for the Blue Economy in Thailand
Investment Report

Analyzing the importance of Thailand's blue economy, this report considers ways to mobilize and scale up private sector investment in areas such as aquaculture and marine renewable energy to help bolster sustainable blue development. It explains why factors such as limited awareness of ocean economy opportunities, a lack of investment-ready at-scale projects, and insufficient monitoring are proving a barrier to investors. The report shows how creating an enabling environment and developing products such as sustainability-linked loans and technical assistance grants could help drive investment flows to the sector and safeguard Thailand's blue economic growth.

About the ASEAN Catalytic Green Finance Facility

The ASEAN Catalytic Green Finance Facility (ACGF) is an innovative finance facility dedicated to accelerating green infrastructure investments in Southeast Asia. It supports ASEAN governments to prepare and source public and private financing for infrastructure projects that promote environmental sustainability and contribute to climate change goals. The ACGF is a facility under the ASEAN Infrastructure Fund, owned by ASEAN member states and the Asian Development Bank, which also administers the facility.

About the Asian Development Bank

ADB is committed to achieving a prosperous, inclusive, resilient, and sustainable Asia and the Pacific, while sustaining its efforts to eradicate extreme poverty. Established in 1966, it is owned by 68 members —49 from the region. Its main instruments for helping its developing member countries are policy dialogue, loans, equity investments, guarantees, grants, and technical assistance.

ISBN 978-92-9270-427-8

ASIAN DEVELOPMENT BANK
6 ADB Avenue, Mandaluyong City
1550 Metro Manila, Philippines
www.adb.org

9 789292 704278

www.ingramcontent.com/pod-product-compliance
Lightning Source LLC
Chambersburg PA
CBHW042034220326
41599CB00045BA/7387